# NMR

Basic Principles and Progress

# 28

Editors: P. Diehl E. Fluck H. Günther
R. Kosfeld J. Seelig

# In-Vivo Magnetic Resonance Spectroscopy III: In-Vivo MR Spectroscopy: Potential and Limitations

Guest-Editor: M. Rudin

With contributions by
N. Beckmann, J.S. Cohen, J.R. Griffiths,
J.S. Ingwall, O. Kaplan, R.J. Maxwell,
M.J.W. Prior, M. Rudin, A. Sauter,
S.R. Williams, P.C.M. van Zijl

With 56 Figures

Springer-Verlag
Berlin Heidelberg NewYork
London Paris Tokyo
Hong Kong Barcelona
Budapest

ISBN 3-540-55029-1 Springer-Verlag Berlin Heidelberg NewYork
ISBN 0-387-55029-1 Springer-Verlag NewYork Berlin Heidelberg



Typesetting: Thomson Press (India) Ltd, New Delhi
Offsetprinting: Colordruck Dorfi GmbH, Berlin; Bookbinding: Lüderitz & Bauer, Berlin
51/3020 5 4 3 2 1 0 Printed on acid-free paper

# Foreword

Since the first report on NMR in living cells in 1974, in vivo magnetic resonance spectroscopy (MRS) has evolved into a potent technique for studying metabolic processes in an undisturbed biological system. In principle, MRS should provide well-resolved spectra of a selected region of interest allowing the identification and quantitation of as many metabolites as possible. However, a number of methodological problems have to be solved in order to generate useful MRS data. For example, the spectra should contain only little or no contribution from signals from outside the region of interest. Furthermore, proper quantification of metabolite concentrations is often difficult since the large number of compounds contributing to the spectra produce overlapping resonances. Similarly, the resolution of the spectra is often impaired by variations in tissue magnetic susceptibilities, leading to large linewidths. Many metabolites occur at low concentrations and a poor signal-to-noise ratio is another major obstacle in proper quantification. The use of radiofrequency (RF) probes with an inhomogeneous RF field profile, such as the surface coil, causes problems when defined RF excitation has to be used, e.g. for proper signal inversion.

During the last 10 to 15 years, a number of solutions to the above mentioned problems have been developed and a proper review of the methodological achievements in in vivo MRS has long been overdue. Rather than writing a textbook, we decided to invite the acclaimed experts to review their respective speciality. We did not intend to discuss in vivo MRS from its basics, but rather refer to textbooks and review articles which have covered these aspects. Basic principles e.g. in NMR probe design or data processing are considered as 'prior knowledge', with which the reader should be familiar to some extent.

The book is divided into three volumes. Volume I contains sections on MR probe construction, design of RF pulses, and spectrum analysis. Volume II is devoted to localization techniques, i.e. methods for selecting a region of interest, and techniques for spectral editing. Volume III finally describes potential applications of in vivo MRS. Due to the rapid growth of the field during the last few years it would have been an almost 'infinite' task to provide a comprehensive coverage of all ongoing activities. We thus have asked the respective authors to focus on methodological and conceptional aspects in order to illustrate the potential of MRS more than the breadth of application.

We would like to thank all contributors for their efficient and pleasant collaboration and for keeping within the short deadlines imposed on them by the editors. We are also indebted to Springer-Verlag, in particular Dr. R. Stumpe and Ms. P. Frank, for their assistance in general and the editorial supervision.

Basel, September 1991

M. Rudin
J. Seelig

Guest-Editor

Dr. M. Rudin

Preclinical Research 386/206
Sandoz Pharma Ltd.
Ch-4002 Basel

# Table of Contents

**Table of Contents of Volume 26**

In Vivo Magnetic Resonance Spectroscopy I:
Probeheads and Radiofrequency Pulses, Spectrum Analysis

**Table of Contents of Volume 27**
In Vivo Magnetic Resonance Spectroscopy II:
Localization and Spectral Editing

# In Vivo NMR of Cells and Perfused Organs

Should experiments on cell preparations or perfused organs be called 'in vivo' or 'in vitro'? While some scientists strictly reserve the term 'in vivo' to studies of intact organisms, i.e. animals and humans, we have decided to adopt a somewhat broader view, also denoting cell suspensions and perfused organs as "in vivo". The first in vivo studies were carried out on isolated erythrocytes and, for several years, studies on isolated cells and perfused organs were the only way to collect meaningful in vivo data. This was due to technical limitations, such as small magnet bores and the lack of suitable spectral localization techniques. The methodological developments of the past 15 years, providing magnets with bore sizes of 30 cm and more, of techniques such as topical NMR, surface coils and image-guided volume selection, resolved these problems and have made it possible to study whole animals and humans. Since then, in vivo studies with intact organisms have matured to an independent scientific discipline and have outnumbered studies of cell model systems and organ preparations.

Nevertheless, studies on isolated cells and organs still represent a valuable tool for the understanding of metabolic processes. By carefully controlling the experimental conditions, selected biochemical mechanisms can be studied in much greater detail than would be possible in the intact organism. Typical applications are the study of cellular energy metabolism, of substrate turnover, of phospholipid pathways or membrane transport processes. The following chapter by Kaplan, van Zijl and Cohen reviews methodologies and applications of NMR studies of isolated cells and perfused organs.

# NMR Studies of Metabolism of Cells and Perfused Organs

**Ofer Kaplan, Peter C.M. van Zijl, and Jack S. Cohen**

Pharmacology Department, Georgetown University Medical School, Washington DC 20007, USA

## Table of Contents



This chapter documents the application of NMR spectroscopic methods to studies of the metabolism of cells and perfused organs. Progress in this area is attested to by the sheer number of publications. This area of research stands alone, as a means of shedding light on the metabolism and pharmacology of cells and organs, but also exists as a basis for the applications of NMR spectroscopic methods to in vivo and clinical situations. Methods of obtaining cellular extracts are included, but the bulk of the chapter is devoted to the methods that have been developed to obtain spectra from intact cells perfused in the NMR tube. Such methods as embedding in gels, use of microcarrier beads, hollow fibers, and spheroids are described and compared. Most of this work involves $^{31}$P-NMR, but some $^{13}$C-NMR tracer studies are also reported. A section is also devoted to the consideration of proton NMR as a means to monitor metabolism, and the necessity of not only suppressing the water signal, but also of removing all signals of extracellular origin, which has recently been accomplished. Finally some work on perfused organs is described and collated into a table that could prove useful for the reader. The question arises as to the relevance of the results with perfused organs relative to those in vivo. As clinical spectroscopy develops, future work must address this question.

NMR Basic Principles and Progress, Vol. 28

# 1 Introduction

The fact that magnetic resonance techniques have become important in clinical imaging is well known. Perhaps less well appreciated is the role of NMR as a research tool to study the metabolism of isolated intact cells and organs. This method can provide important information on biochemical processes, and can be used for the identification of signals in in vivo spectra.

In relation to the subject of this article, it is necessary to define some terms. Studies with intact cells or organs cannot be termed "in vivo," since they are carried out outside the live body. Although these studies have also been called "in vitro", this terminology is in contrast to the common use of in vitro to mean for instance, studies on purified enzymes or extracted metabolites. It is far preferable to study metabolic events in the intact cellular milieu. This chapter focusses on data derived from ex vivo studies of cultured cells, and experimental perfused organs, excluding clinical studies or analysis of materials obtained by biopsies.

Usually for meaningful results these studies are carried out with large numbers of cells, or with whole perfused organs. Clearly the relevance to the in vivo situation of such "model" systems must be proven in each case, and sometimes the results are found wanting. However, for the understanding of metabolic pathways, and the ability to control the experimental conditions, such ex vivo NMR studies are a very valuable approach. In some cases they provide the only noninvasive means of studying certain pathways that previously could be investigated only in solutions of extracted metabolites or purified enzymes, the mechanisms of which are clearly altered by the lack of conditions that simulate the natural cellular environment.

Perfused intact cells represent perhaps the best approach to the non-invasive study of metabolism. In contrast to the in vivo situation, these cells are homogeneous, particularly when grown in culture conditions. Moreover, they are metabolically stable, and thus much more advantageous to study than suspensions of cells. Therefore, the bulk of the chapter will be devoted to studies of perfused intact cells using $^{31}$P-NMR spectroscopy. Generally, these studies provide information on normal cellular energetic status [1–7] substrate utilization and metabolic pathways [8–10] phospholipid pathways [11, 12], pH changes [13–15] and membrane permeability and ion and water distribution [16]. Using these methods, differences between cell lines can be delineated [17, 18], and the effects on metabolism following manipulation with nutrients [6, 13, 19–21], hormones [22, 23], drugs [24–27], growth factors [18, 28], and hyperthermia [29] can be monitored. In this review we discuss the methods used for perfusing intact cells, and provide examples of applications of metabolic studies done in each case. Most applications are with $^{31}$P, $^{1}$H, $^{13}$C, but $^{23}$Na and $^{19}$F studies will also be mentioned.

This review focuses on techniques rather than each NMR nucleus. However, proton NMR may play an important role in NMR of intact cells, mainly due to the inherent high sensitivity and universality of this nucleus. This should allow acquisition of data with sufficient signal-to-noise within a minute, thus enabling the monitoring of metabolic processes on this time scale. However, due to technical difficulties in suppressing the dominating water line, the potential of $^{1}H$ NMR has been less realized than that of phosphorus. Most proton studies have been peformed on erythrocyte suspensions, work initiated by Campbell et al. [30, 31]. Water suppression was successfully attained by combining spin-echo approaches [32–34] and the strong reduction of the spin–spin relaxation time ($T_2$) for the exchanging water protons as a function of magnetic field strength [35]. When shorter echo times ($T_E$) are required, saturation transfer experiments [36] can be used. Techniques for suppression of unwanted signals have been reviewed [37–41] and so have most applications to erythrocyte suspensions [39, 42–44].

Intact viable tumor cells were first studied by $^{1}H$ NMR by Agris and Campbell [45], who demonstrated that the phosphocholine concentration increases during dimethylsulfoxide-induced differentiation of Friend Leukemia cells. Navon et al. [46] studied differentiation of hybrid neuroblastoma/glioma cells induced by prostaglandin E1 and theophylline. The undifferentiated cells contained significantly higher levels of glutamine, an observation supported by cell extract data. Irving et al. [47] monitored the metabolism of $^{13}C$-labeled ethanol with proton NMR. Reglinski and co-workers studied glycolytic patterns in HeLa cells [48] and in the leukemic cell line J-111 [49] by monitoring lactate. In the latter study, treatment with pharmacological concentrations of the drug doxorubicin caused enhanced anaerobic glycolysis and cell death.

The most detailed study of the features of proton NMR spectra of intact cancer cells has been made by Mountford and coworkers [50, 51]. For the rat adenocarcinoma cell line R13762, they found that 35% of the protons contributed to the high resolution spectrum. Of these "long-$T_2$" protons, about 20% is concentrated in the methylene resonance around 1.3 ppm [52]. It was also found that the intensity of this resonance increases with increasing neutral lipid content of the cells for a range of embryonic, malignant, and transformed cells [53], but that unstimulated lymphoblasts and red blood cells do not generate this high-resolution lipid spectrum [51]. Earlier 2D-NMR work had assigned this resonance to be mainly trigyceride which is not in bilayer form [54]. Since neutral lipids are commonly associated with lipoprotein particles, experiments on serum lipoproteins were initiated [55]. It was found that the COSY spectrum of the intact cells [56] closely resembled that of a malignancy associated lipoprotein (MAL) [57], which has been reported [58, 59] to be shed by malignant cells. This is of interest, since previous experiments had related the methylene resonance to metastasis and drug resistance [57], which may well be a function of membrane composition. Also, reports by Poste and Nicolson [60] had indicated that vesicles shed by metastatic cells can transfer metastatic potential to nonmetastatic malignant cells. Based on their series of proton NMR experiments and supported by chemical analysis, Mountford et al. [51] have proposed a new model for the organization of lipids in membranes of malignant and stimulated cells. In this model neutral lipid domains containing mainly triglycerides are intercalated in the plasma membrane, and their relative contribution to total lipid increases in malignant and stimulated cells.

Since NMR is an insensitive method, it is necessary to have large members of cells within the coil ($10^7$–$10^9$), and this somewhat limits its use to cell lines that can be grown readily in culture conditions. Not surprisingly, most of the studies of cell metabolism have focussed on cancer, which is an area of major biomedical interest. Moreover, many immortal human, as well as experimental, cancer cell lines have been established, so they can be grown in the large numbers necessary for NMR experiments.

Normal cells are more difficult to grow, and perfusion NMR studies of their metabolism are less common. Notably, cells derived from the blood system, that can be obtained in large quantities, such as red blood cells [61, 62] and lymphocytes [15, 63] have been used. Other studies involve retinal cells [64], kidney tubules [64, 66], adrenal medullary cells [5], macrophages [67,], fibroblasts [1, 16, 19, 68], mammalian ovarian cells [13], as well as plant cells [69–71], microorganisms [4, 72], and protozoa [19].

With regard to perfusion of isolated organs, since so many applications have been made to heart and a few other organs, we have grouped many of them together in a table. This is not intended to be a comprehensive survey, the relevance of the results obtained to the working organ in vivo is a major consideration in these studies.

## 2 Metabolism of Cells

### 2.1 Cellular Extracts

Data obtained from cellular extracts are essential for the interpretation of results of NMR studies of intact cells and tissues. Therefore, although these are not metabolic studies of intact cells, they should be considered together with studies of cells and organs.

The advantage of studying extracts are that the improved resolution enables assignments of proximate signals with the aid of pH titration curves of the chemical shifts, coupling constants, and additions of known compounds to the extract solution [73]. In contrast to living cells, there are no time constraints, and with prolonged data accumulation compounds present at low concentrations may be observed and quantitated [74–77]. Intact cells are in aqueous solution, and until recently this hampered the applicability of proton NMR to studies of cells; this problem is not encountered when studying extracts, since they are usually lyophilized and are redissolved in $D_2O$.

However, data obtained from extracts should be evaluated with caution, since they can represent artifacts due to the extraction procedure, and may be misleading. Generally, the problems associated with NMR studies of extracts are: a) compounds which are observed in intact cells may not pass into extracts due to insolubility [78]; b) the extraction procedure and agents may cause molecular changes in cellular metabolites [79]; c) there may be differences between cellular metabolites in susceptibility to the extraction procedure, or in extraction yield, thus leading to errors in quantitative measurements and comparisons; d) there may be a loss of cellular material during the extraction process; e) even assignments of $^{31}P$ signals by pH titrations may lead to confusion due to similar titration curves for different compounds. This can be addressed by using selective heteronuclear $^{1}H$ decoupling [77], or 2D $^{1}H$-$^{31}P$ correlation spectroscopy [78].

In most NMR studies perchloric acid extractions, containing water soluble metabolites [80, 81], have been used. The final concentrations of perchlorate are of critical importance; values between 0.5–1.7 M have been reported [20, 75, 77]. It was demonstrated that the optimal concentration to minimize artifacts is 0.5 M [79]. In order to prevent chemical changes, the perchlorate should be neutralized as quickly as possible, the solutions should be kept cold, and extracts should be dried and stored frozen. It is often necessary to remove metal cations from extracts by purification, or passing through chelating agents, in order to reduce line broadening [73, 75, 82]. In several instances a modified extraction method, as described by Veech et al. [83], was used [84, 85], in which the tissue or cells are mixed with cold 0.1 N HCl in absolute methanol, and after 10–15 min the perchlorate is added.

Differences between extraction of tissues and cells are noteworthy; tissues should be frozen in situ (usually by freeze clamps in liquid nitrogen), and mechanically fragmented and powdered, before mixing with perchloric acid. These potential harmful steps are not mandatory in cellular extraction. However, freezing of cells (yeasts) with liquid nitrogen prior to extraction with perchloric acid is sometimes beneficial [86]. Moreover, in cellular extracts, signals arising from connective tissue elements and blood vessels, present in extracts of tissues, are avoided.

Initially, $^{31}P$-NMR studies of extracts of tissues (muscle) were investigated [74, 87]. Subsequently, the profiles of metabolites of many organs were described using also $^{1}H$ and $^{13}C$-NMR. Many studies of extracts focussed on brain [75, 84, 88–94] and metabolic sequences of pathological states like hypoglycemia [95], ischemia [96], and neoplasia [97]. This can be partly explained by the lack of experimental models of perfused brain. $^{31}P$, $^{1}H$, and $^{13}C$ nuclei were also used in studying the extensive metabolic activity of the liver [98–100], and the effects of ethanol [101]. Other organs of which the extracts have been studied by NMR include the heart [102, 103], kidney [85, 104], pancreas [105, 106], muscle [107], eye [108, 109], reproductive organs [110], and normal and neoplastic breast tissues [111]. Methods of in vivo NMR spectroscopy are now much improved, and extracts are used today mainly to confirm and quantitate in vivo results.

The first $^{31}P$-NMR studies of transformed cells were reported by Navon et al. [73]; extracts were used to assign metabolites in Ehrlich ascites tumor cells (see below—cellular suspensions). Further studies demonstrated substantial qualitative and quantitative differences in $^{31}P$ spectra of various mammalian cell lines [112], including human Hela cells [112, 113].

Identification of $^{1}H$, $^{31}P$, and $^{13}C$ resonances in perchloric acid extracts of murine fibrosarcoma tumor were performed by Evanochko and co-workers [77]. The findings of this important work were used in many subsequent studies for signal assignments, e.g., in human cancer cell lines, such as breast carcinoma [17] melanoma [114], retinoblastoma [21], and colon carcinoma [20]. In the latter study $^{13}C$-NMR of cellular extracts, using $^{13}C$-labeled glucose and acetate, provided information on flux through glycolytic pathways. It was determined that differentiation is correlated with a large increase in the activity of oxidative metabolism [115]. Phosphomonoester and phosphodiester compounds have recently attracted much interest regarding their roles in cellular proliferation and differentiation. $^{31}P$-NMR of extracts are now widely used to investigate these multiple signals, which are poorly resolved in vivo.

Studies of extracts of Morris hepatoma, as well as of cultured cells derived from it, demonstrated an unusual $^{31}P$ resonance at 16.5 ppm [76]. Both perchloric acid and aqueous phase of chloroform/methanol/water extracts (see below) were done. The signal was identified as *myo*-inositol 1,2-(cyclic) phosphate, and was absent from normal tissues. The significance of the accumulation of this compound in transformed cells is still unclear, and attempts to discover biochemical "fingerprints" of malignancy by NMR have been fruitless. However, proton NMR spectra of perchlorate extracts of human colon adenocarcinoma cells showed significant amounts of a related compound, *myo*-inositol [115].

Szwergold et al. [14], identified D-sorbitol 3-phosphate (at 6.5 ppm in $^{31}P$-NMR spectra) in perchloric acid extracts of lenses, and found manyfold increase in its concentrations in lenses of diabetic rats. The identification was confirmed by GC-MS. Also, $^{1}H$ NMR combined with GC-MS techniques provided quantitative data on metabolites in carrot root [89], and $^{31}C$ NMR and GC-MS contributed to the knowledge on glycogen pathways in the liver [116]. $^{31}P$-NMR of extracts together with HPLC were used in analysing the nucleotide compounds in RIF-1 tumors [77]. Combination of NMR studies of extracts with other analytical methods seems to offer the benefits of providing complementary data, that cannot be obtained by either method alone.

Combined analysis of both $^{1}H$- and $^{31}P$-NMR spectra of extracts of drug-sensitive and multidrug-resistant MCF-7 human breast cancer cells enabled quantitative comparisons between the concentrations of major metabolites, as well as of their precursors (Fig. 1) [117]. $^{31}P$-NMR studies of intact cells had demonstrated striking spectral differences between the two cell types [17]. However, since the number of compounds detectable by $^{31}P$ MRS is limited, the

**Fig. 1 a, b.** Perchloric acid extracts of wildtype and multidrug-resistant MCF-7 breast cancer cells; (**a**) $^{1}H$-NMR spectra at 400 MHz in $D_2O$, line broadening, 1 Hz; (Ch, choline; TSP, trimethyl-silylpropane sulfonic acid); (**b**) $^{31}P$-NMR spectra at 162 MHz, line broadening, 3 Hz; Ref. is trisodium trimetaphosphate [117]. Reproduced by permission

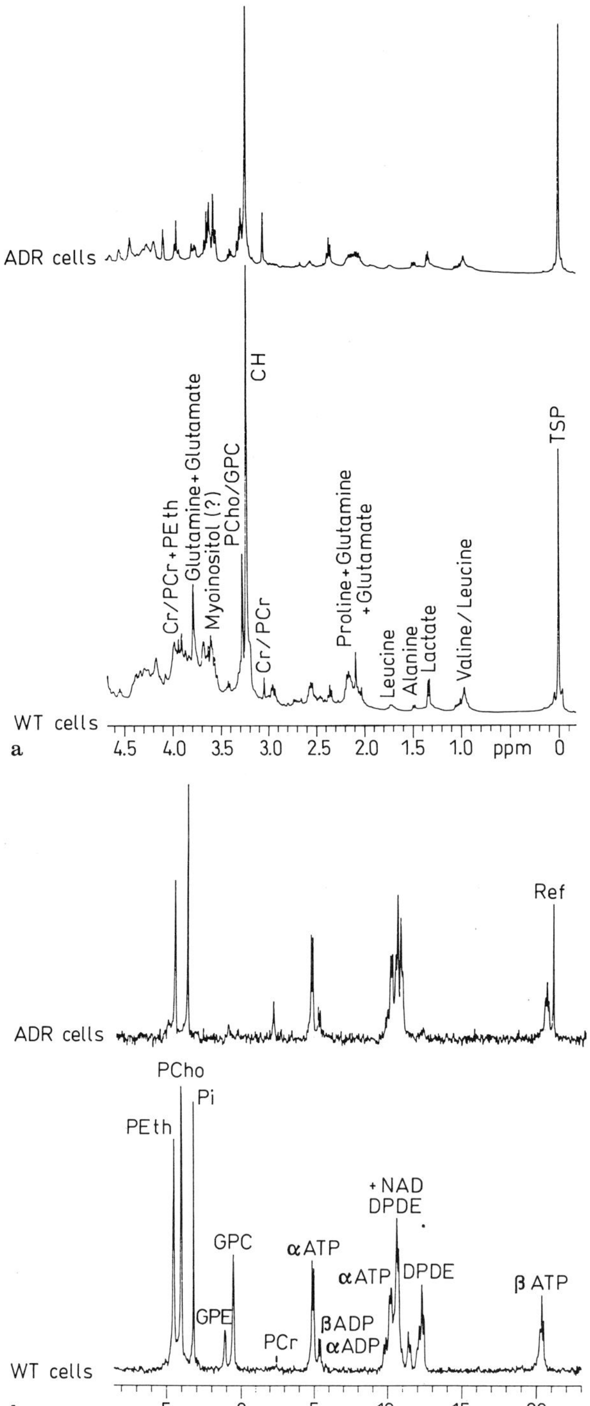
ADR cells
WT cells
Cr/PCr + PEth
Glutamine + Glutamate
Myoinositol (?)
PCho/GPC
CH
Cr/PCr
Proline + Glutamine
+ Glutamate
Leucine
Alanine
Lactate
Valine / Leucine
TSP
4.5 4.0 3.5 3.0 2.5 2.0 1.5 1.0 ppm 0
a
ADR cells
Ref
WT cells
PCho
Pi
PEth
GPC
GPE
PCr
α ATP
β ADP
α ADP
+ NAD
DPDE
DPDE
β ATP
5 0 -5 -10 -15 ppm -20
b

nature of these differences could not be fully evaluated. Also proton NMR by itself was insufficient, due to the fact that $^{1}H$ NMR signals of some phosphorus-containing compounds and their precursors resonate at almost identical chemical shifts. Only combined analysis of $^{31}P$ and $^{1}H$ NMR of extracts could provide information on the mechanisms of multidrug-resistance.

Metastatic potential of tumor cells was studied in chloroform/methanol treated rat mammary adenocarcinoma cells [118]. This is a lipid extraction method [119], and it was used in order to improve resolution in the lipid region of proton spectra, and to demonstrate metastatic-associated membraneous-bound glycoproteins. However, the origins of the spectral differences between tumor cell subpopulations were not identified. Recently, in an elegant in vivo NMR study of rat fetal brains, lipid extracts were used to demonstrate membranous phospholipid profiles in $^{31}P$-NMR, and perchloric acid extracts to assign in vivo detected signals by $^{1}H$ and homonuclear 2D NMR [94]. However, the information obtained from lipid extracts is significantly reduced, and the role of this method in NMR studies of metabolism seems to be rather limited.

Studies of extracts have had a principal role in providing the basic knowledge on the common $^{31}P$-, $^{1}H$-, and $^{13}C$-NMR signals of mobile metabolites of cells, and this data is routinely used now. Extracts are still useful in well-defined circumstances, especially for assignments of poorly resolved in vivo signals, for quantitation in combined studies of several nuclei, and with other analytical modalities.

## *2.2 Cellular Suspensions*

NMR studies of cell suspensions have furnished valuable information on cellular biochemistry and physiological processes (see reviews 120–123). There is a vast literature on these studies, which is beyond the scope of this review. Moreover, cellular perfusion techniques suited for NMR experiments, introduced and elaborated in the last decade, are now much preferred for studies of metabolism, and are the focus of our discussion (see below).

Due to NMR's inherent low sensitivity, cellular studies can be performed only at high field and with dense cellular populations. Packed cell suspensions in NMR tubes experience shortage of oxygen and nutrients, accumulation of waste toxic products, and pH changes. Metabolic studies often include chemical manipulations which require the capability for reversible environmental changes; cellular suspensions are clearly inferior to perfusion experiments in this regard.

Another drawback of studies of cell suspensions is that for most cell types this is not their normal physical state. Some cell types are more suitable for suspension studies; namely blood elements which are normally in suspension, and have a low baseline metabolic activity so they can withstand suspension conditions better than other cells, and also microorganisms. However, even these cells are highly packed in the NMR tube, and in these stagnating conditions the natural metabolic steady state cannot be maintained. The only exception is perhaps anaerobic microorganisms.

Sedimentation of suspended cells aggravates their metabolic status, and may lead to erroneous data, since a substantial portion of the cells is likely to be below the NMR coil. Several techniques have been suggested for overcoming these obstacles. The simplest, and perhaps the most useful for both problems, is to bubble the cells with air, oxygen, or 95% $O_2$/5% $CO_2$. This of course affects field homogeneity, but not to a critical extent for most cellular studies, especially when small bubbles are used. Gating the bubbling with signal acquisitions was proposed as a means to overcome this problem [73], and in other cases it was necessary to augment oxygenation through additional tubes above the NMR coil [124]. Other techniques include stirring with motor-driven propeller [125], improvements of gas delivery devices [126, 127], and addition of polysaccharide arabinogalactan to increase the density of the suspending buffer [128]. Also, most studies using cellular suspensions were conducted at ambient temperature, thus alleviating metabolic instability; however, it is clearly preferable to study metabolism at 37 °C.

Contrary to a common notion, $^{13}C$, and not $^{31}P$-NMR was used in the first reported studies of intact cells in suspensions. Matwiyoff and co-workers investigated blood cells with $^{13}C$ labeled $CO_2$, CO, and $CN^-$ ion [129], and yeasts with [1-$^{13}C$]-glucose [130], and followed their metabolism. The first NMR data on a natural abundant nuclide in intact cells was described by Moon and Richards [131], who recorded $^{31}P$-NMR of erythrocytes and found that the chemical shifts of some of the intracellular phosphates are sensitive to intracellular pH. Shortly thereafter, Henderson et al. [132] found a reversible decrease in 2,3-diphosphoglycerate in $^{31}P$-NMR spectra of stored erythrocytes. These studies presented, perhaps for the first time, a method to observe the intracellular milieu of intact metabolizing cells, and launched an extensive field of research. The first $^{31}P$-NMR studies of malignant cells were those of Navon et al. [73]; glycolysis pathways and pH changes were investigated following conversion from aerobic to anaerobic conditions, and administration of deoxyglucose, in Ehrlich ascites tumor cells (Fig. 2).

The activity of the Yale (previously Bell Laboratory) group in studying the metabolism of suspended cells by $^{31}P$- and $^{31}C$-NMR techniques is noteworthy. It included investigating energy metabolism, carbohydrate pathways and pH in bacteria [73, 133], yeasts [86, 134], liver cells [135], and tumor cells [73, 112], to name a few.

$^1H$ NMR of intact cells has attracted much interest [38, 45, 49, 50], most probably due to the high inherent sensitivity and universality of this nucleus. Since proton NMR of cells encounters major experimental difficulties it is discussed in a special section of this review (Sect. 2.4). NMR of $^{23}Na$ [136], $^{39}K$ [137], $^{19}F$ [138], $^{15}N$ [139] nuclei were also applied to studies of cellular suspensions.

There are now enough methods to restrain cells in an NMR tube and perfuse them that an appropriate method can be adjusted to almost all cell types and experimental requirements (next section). Therefore, it is highly recommended that metabolism is studied by NMR in perfused cells. NMR studies of

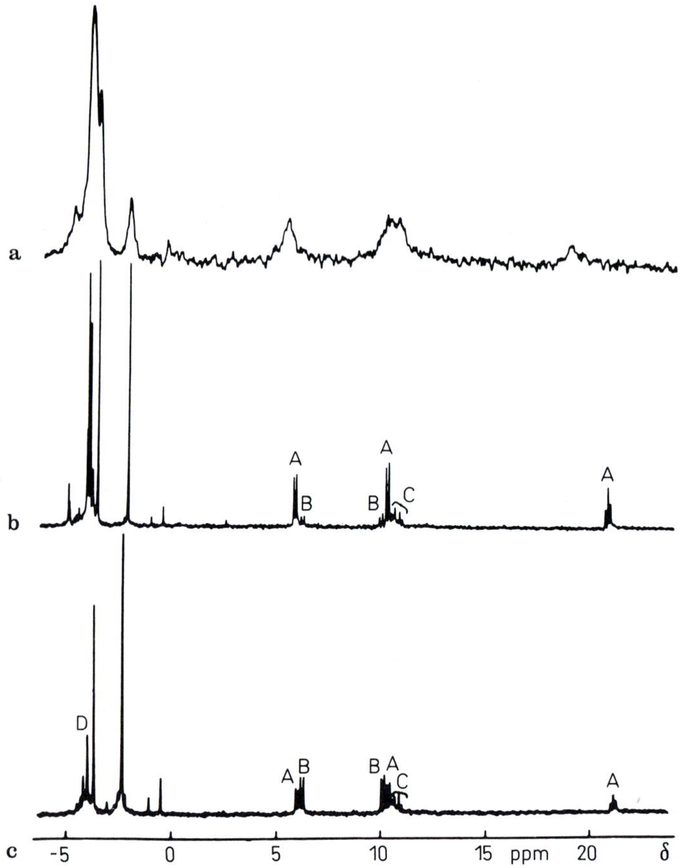


**Fig. 2 a–c.** $^{31}$P-NMR spectra at 145.7 MHz of (**a**) a suspension of Ehrlich ascites cells; (**b**) and (**c**) are perchloric acid extracts at pH 6.7 and 6.9, respectively (AB, and C, are ATP, ADP, and NAD, respectively) [73]

suspensions of mammalian cells are warranted in a limited number of experimental circumstances, particularly using scarce compounds or isotopically labeled substrates [140], when only short-term experiments are performed, and when extracellular metabolites are of interest (Fig. 3).

## *2.3 Perfusion of Intact Cells*

One of the most advantageous features of NMR as a tool for studies of metabolism is the capability to continuously follow biochemical events, and to monitor physiological processes in intact cells. For prolonged experiments cellular suspensions are not adequate, especially whenever one attempts to

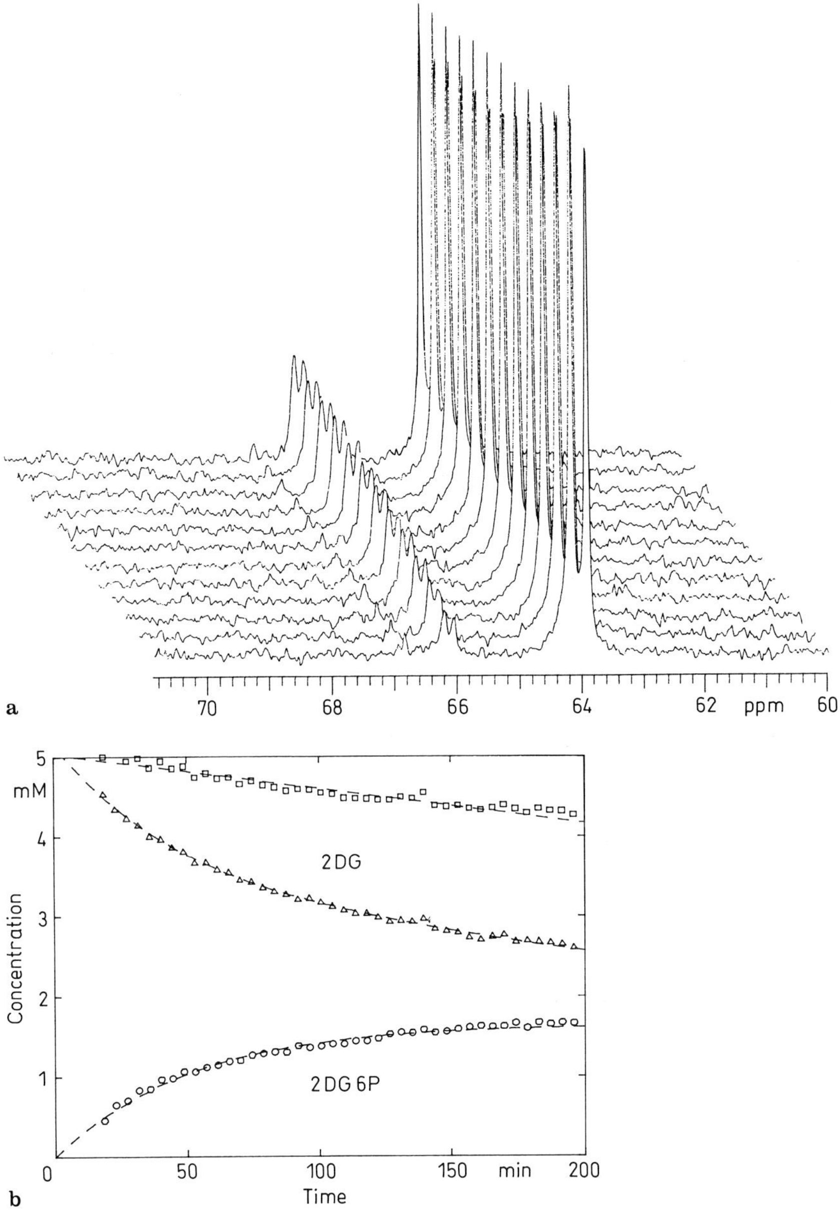


**Fig. 3a, b.** Metabolism of $^{13}C_6$ 2-deoxyglucose (2DG, *right peak*) to $^{13}C_6$ 2DG6P (*left peak*) in a suspension of MCF-7 cells; (**a**) $^{13}C\{^1H$ decoupled$\}$-NMR spectra at 100 MHz; (**b**) Plot of time course showing exponential decrease in 2DG (*triangles*), exponential increase in 2DG6P (*circles*), and total linear decrease (*squares*) [140]. Reproduced by permission investigate the effects of metabolite derivatives, hormones, drugs, irradiation, etc. It is imperative that the cells are continuously perfused and furnished with nutrients, and the waste products of metabolism are removed, in order to maintain the cells in a proper metabolic state. Perfusion cannot be performed in cells freely suspended in the NMR tube, since the flow would wash them away, or the cells would block filters if used; therefore, the cells should be restrained. Over the last decade several methods of cellular restraint and perfusion inside the NMR tube have been developed. Some general differences between the various methods should be emphasized; some are appropriate only for anchorage-dependent cells, while in others also anchorage-independent cells can be studied; cells sequestrated in some of these methods do not multiply, and thus studies of cellular growth cannot be performed; the technical aspects and difficulties in maintaining experimental conditions are markedly variable.

In all experiments that are expected to last for several days sterility must be maintained, and this should be considered in selecting the appropriate perfusion method for a certain study. This can be accomplished by either using large capacity filters, or closed systems. It should be noted that recirculating perfusion is adequate as long as the nutrients are not depleted, and the waste products of metabolism are not accumulated. Moreover, when labeled isotopes are used, recirculation perfusion is often mandatory due to the scarcity and cost of these agents. However, for prolonged experiments, especially whenever normal cellular growth is necessary, it is recommended to perfuse with fresh medium. It is essential that all metabolic studies should be performed at the physiologic temperature of 37 °C.

Various methods are discussed here, with details regarding technical aspects and experimental procedures, to facilitate researchers in choosing the most appropriate method. References to the applications of these methods in studies of cellular metabolism will be presented. The two techniques, which are most widely used are bead attachments and embedding in agarose threads.

### *2.3.1 Microcarrier Beads*

The first technique of cell perfusion for NMR studies was described by Ugurbil et al. [1], and was based on attaching anchorage-dependent cells to dextran microcarrier beads (Cytodex). Cells were seeded on beads in the incubator, attached and multiplied, and coated the beads as monolayers. The beads were then transferred to the NMR tube, which was placed in the spectrometer with a special transmitter-receiver coil (four-turn solenoid) around it. The tube was fitted with removable Plexiglas end-pieces with central openings. On the inner surface of one of the end-pieces, a nylon mesh screen was mounted. The perfusion apparatus consisted of a reservoir filled with perfusion medium, a peristaltic pump, a gas exchanger made of gas-permeable silistic tubings enclosed in a chamber flushed with 5% $CO_2$/95% $O_2$ air, and the effluent was directed back to the reservoir. The early $^{31}$P-NMR studies by Ugurbil et al. [1] demonstrated different effects of oxidative phosphorylation and glycolysis inhibitors on transformed and "normal" fibroblasts.

Besides dextran, beads of other compositions and sizes are now available (Fig. 4): e.g. treated polystyrene (Biosilon or Cytosphere, 160–300 $\mu$m diameter) polyacrylamide (BioCarriers, 120–180 $\mu$m); agarose polyacrolein (Acrobeads). Gelatin or dextran beads covered by a layer of denaturated collagen (Cytodex 3) are enzymatically soluble, which facilitates cell recovery. Pretreatment of Biosilon beads with poly-L-lysine was reported to improve cell attachment to 95% [6]. Smaller beads show the benefit of reducing the "dead" space in the NMR coil region, and improving S/N [141]. The volume of beads and numbers of cell used in the published studies varied between 2.5 ml with $2–6 \times 10^7$ cells in 10-mm NMR tube [22], 25–30 ml with $1.5–4 \times 10^8$ cells [14], and $2.5–3 \times 10^8$ cells in 20-mm tube [143]. Perfusion rates ranged from 1.5 [22] to 15 ml/min [143], and in most experiments the perfusate was recirculated after being reoxygenated in the reservoir.

Most of the ensuing NMR studies employing microcarrier beads investigated metabolism of cancer cells. The effects of epidermal growth factor (EGF) on phosphorus metabolism of A-431 human epidermoid carcinoma cells, and SV-40 virus transformed mouse fibroblasts were investigated by Melner et al. [28]. EGF induced late ATP depletion in perfused A-431 cells concomitant with elevation of sugar phosphates. Similar changes were noted in the perfused transformed fibroblasts, but at different rates. Glynn et al. [144] studied phospholipid (PL) metabolism in neutroblastoma cells by $^{31}$P-NMR, and found that the undifferentiated cells were characterized by low signals of PL precursors (PEth-phosphoethanolamine, and PCho-phosphocholine), and that these signals increased when the cells were induced to differentiate. These reports demonstrated the applicability of this method for studying metabolic processes of cellular stimulation by autocrine and hormonal agents, and differentiation.

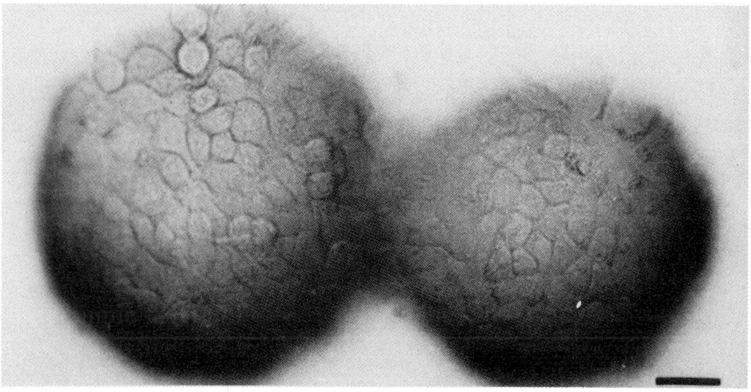

**Fig. 4.** Human colon adenocarcinoma cells (HT29) grown on Biosilon microcarrier beads (bar is 50 $\mu$M) [143]. Copyright © 1987; reprinted by permission of Wiley–Liss, a division of John Wiley and Sons, Inc.

Szwergold et al. [14] investigated the theory that mitogenic stimulation of mammalian cells is followed by intracellular alkalinization. Using 3T3 cells attached to Biosilon microcarrier beads, they found in $^{31}$P-NMR studies, that mitogen induces a minimal pH elevation, which could be abolished when the cells are grown in medium containing physiological concentrations of bicarbonate. These experiments demonstrated that an increase in intracellular pH is not mandatory for cellular transformation. The problem of determining intracellular pH in 3T3 cells was also addressed by Civan et al. [16] using the microcarrier bead method. They showed that perfusing the cells with 2-deoxyglucose, and analysing the signals of its phosphorylated product (2-DG-6-Phosphate) can provide the correct pH. By manipulating the ionic composition of the perfusion solution, the authors gathered information on the operation of a Na/H antiport in the plasma membrane, and demonstrated that the pH was partly regulated by a gradient in $Na^+$ concentration across the membrane.

Neeman and coworkers studied the metabolism of T47D human breast cancer cells. The effects of 17$\beta$-estradiol versus tamoxifen were investigated by both $^{31}$P, and $^{13}$C-NMR. $^{31}$P-NMR data showed that PCho and nucleoside diphosphate increased in cells exposed to tamoxifen, compared to estrogen-treated cells [22]. However, these authors used significantly smaller samples of cells compared to the other groups who use microcarrier beads, and the reported differences in the metabolites levels seem to be rather small. Glucose utilization and glycolysis products were monitored by $^{13}$C-NMR using [1-$^{13}$C]-glucose. Estrogen administration was followed by elevated rates of lactate and glutamate production. The enhancement of glucose metabolism induced by estrogen was found to be inhibited by actinomycin D or cycloheximide, which are mRNA inhibitors [25]. It was postulated that such studies may contribute to understanding the response of human breast cancer to hormonal manipulations.

Cozzone and co-workers used the microcarrier bead method to study metabolism of C6 glioma cells by $^{31}$P-NMR [6], and HT29 human colon adenocarcinoma by both $^{31}$P [143] and $^{13}$C [23] NMR. They monitored the energetic changes in glioma cells following glucose starvation, characterized by depletion of phosphocreatine (PCr) and ATP, concomitant with elevation of inorganic phosphorus levels, with reversal changes upon reperfusion with glucose. They noted an elevated pH compared to normal brain, (details about buffering the perfusion medium and its bicarbonate content were not given—see above Ref. [14]), and a decrease in pH during glucose starvation. In the colon cancer cells, abnormalities in carbohydrate metabolism were studied using [1-$^{13}$C]-glucose. HT29 cells were able to utilize glucose through glycolysis while storing glucose as glycogen. Vasoactive intestinal peptide induced only a transient decrease in glycogen formation.

The microcarrier bead method was also used to study the partition of ATP in the cytosolic and vesicular compartments of adrenomedullary chromaffin cells [5]. Bovine chromaffin cells were isolated, and attached to sulfonated polystyrene beads. Perfusion with oxygenated medium enabled detection of distinct ATP signals from the two different compartments, their quantitation, and the measurement of their relaxation times and molecular mobility. It was concluded that the intravesicular ATP is incorporated within a catecholamine-storage complex.

### *2.3.2 Gels*

The gel methods differ from the microcarrier bead method in two fundamental features: (a) in the gel methods the cells are inside the matrix, and therefore the porosity of the matrix and the ease of diffusion of nutrients are of critical importance; (b) since attachment to the gel is not essential, both anchorage dependent and -independent cells can be studied.

#### 2.3.2.1 Agarose Threads

This method was introduced by Foxall and Cohen [2, 3] and is based on the properties of low-temperature gelling agarose (SeaPlaque), that allows mixing of cells with liquid agarose at 37 °C, and solidification of the mixture at a lower temperature. 1–1.2 ml of cell pellet ($2 \pm 0.5 \times 10^8$ cells) are mixed with equal volume of 1.8% liquid agarose in phosphate-buffered saline, and put in the bath at 37 °C for 5–7 min. The mixture is extruded under low pressure through 0.5 mm internal diameter tubing in an iced bath, into a 10-mm NMR tube containing growth medium (Fig. 5). The solid threads which fill the tube are concentrated without compression at the bottom of the tube, by a plastic insert with the perfusion fittings. The inflow tube, made of Teflon, is 0.5 mm internal diameter and placed near the bottom of the tube, and the outflow is directed into openings in the insert, and then into an outflow tube. Perfusion rates (0.5–1.5 ml/min) are maintained by a peristaltic pump, and since the Teflon tubes are permeable to air, it is not required to include a gas exchanger in the perfusion system. Perfusate can be recycled through a pH-stat, when $^{13}C$ or scarce materials are perfused, or can be wasted and the cells are then continuously perfused with fresh medium. A debubbler should be inserted into the apparatus prior to the insert, to remove air bubbles which may spoil magnetic field homogeneity and affect the homogeneity of perfusate flow. In general, the perfusion solution should be the buffered growth medium that is most appropriate for the cells studied, but whenever one needs to accurately monitor the intracellular inorganic phosphorus signal, balanced buffer solution without phosphate can be used. However, ATP levels gradually decrease when cells are perfused for more than 4–5 h without phosphate, but remain stable for more than 24 h in cells perfused with the growth medium (Kaplan, unpublished results).

It is of utmost importance to make very thin threads (0.5 mm diameter). Major differences in diffusion and equilibrium distribution of nutrients were encountered when these threads were compared to 2-mm threads [3, 8, 9]. Using 0.5 mm threads ensures that there is no metabolic compromise, and the cells are viable and in stable energetic status for more than 24 h, while the threads

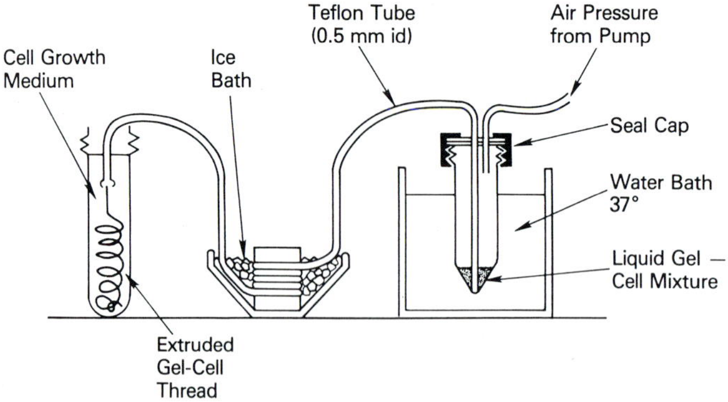


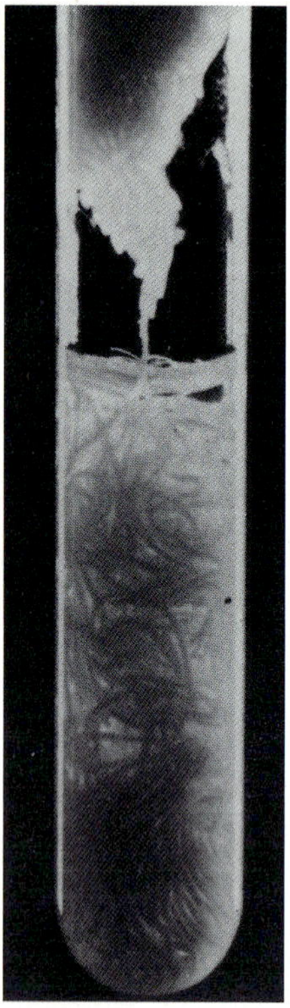

**Fig. 5. a** Apparatus for the preparation of gel threads by extrusion of a cell suspension in liquid gel through a cooled capillary. **b**. Agarose gel threads containing cells; note that the gel threads would be transparent without the cells that make them opaque [3]. Reproduced by permission

maintain their mechanical strength. Moreover, it was found that albumin can readily diffuse into the threads [18] and thus, the effects of polypeptides and high molecular weight compounds can be studied.

The agarose thread method can be applied to studies of metabolism of both microorganisms (yeast, bacteria, protozoa) and mammalian cells. Contrary to preliminary assumptions [145] it is applicable for anchorage-dependent as well

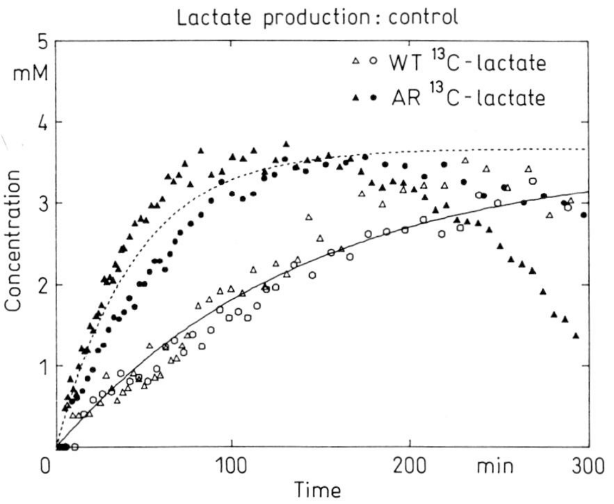


**Fig. 6.** $^{13}$C-lactate production by glycolysis of perfused wildtype (WT; *open symbols*) and multidrug-resistant (ADR; *filled symbols*) MCF-7 cells following addition of $^{13}C_1$-glucose in a closed perfusion system [9]. Reproduced by permission

as anchorage-independent cells. However, small anchorage-independent cells, for example lymphocytes, tend to be gradually washed out of the gel, and for these cells a different matrix, i.e., alginate (see below) is recommended. Most studies with agarose threads have used cancer cells. Thus, the bioenergetics and glycolysis [9] (Fig. 6), phospholipid pathways [11, 12], drug resistance [17, 146], effects of drugs and antimetabolites [26, 147] and growth factors [18] could be effectively investigated in transformed cells.

A modification of the perfusion apparatus was suggested by Geoffrion et al. [149], where the perfusion tubings were thermally isolated by a temperature-controlled water jacket made of large-diameter plastic tubings. This system was also used for studies of perfused rat liver which require high perfusion rates. These authors used their apparatus to monitor the phosphorylated nucleotide pool of murine malaria parasites, embedded in agarose threads, by $^{31}$P NMR, and the lipogenic activity by $^{13}$C NMR, with [12-$^{13}$C]-acetate as a precursor. However, in cellular studies, perfusion rates are low enough that the desired temperature can be achieved in the magnet by using its thermostat without the necessity to control the temperature outside the magnet. Therefore, this modification is often superfluous, and indeed has not often been used.

The agarose thread method has been used with $^{31}$P NMR to investigate phospholipid metabolism of breast cancer cells. Most notable are the effects of substrates and inhibitors of a specific enzymatic process, such as ethanolamine kinase, which in the presence of extracellular ethanolamine resulted in the formation of phosphoethanolamine (Fig. 7A), while a specific inhibitor of choline kinase, hemicholinium-3 resulted in the reduction of the concentration of phosphocholine (Fig. 7B) [11, 12]. Such applications shed light on the synthetic pathways of phospholipid biosynthesis, including the fact that the second step of the three step process, that involving choline-phosphate cytidylyltransferse, is the rate limiting step [12].

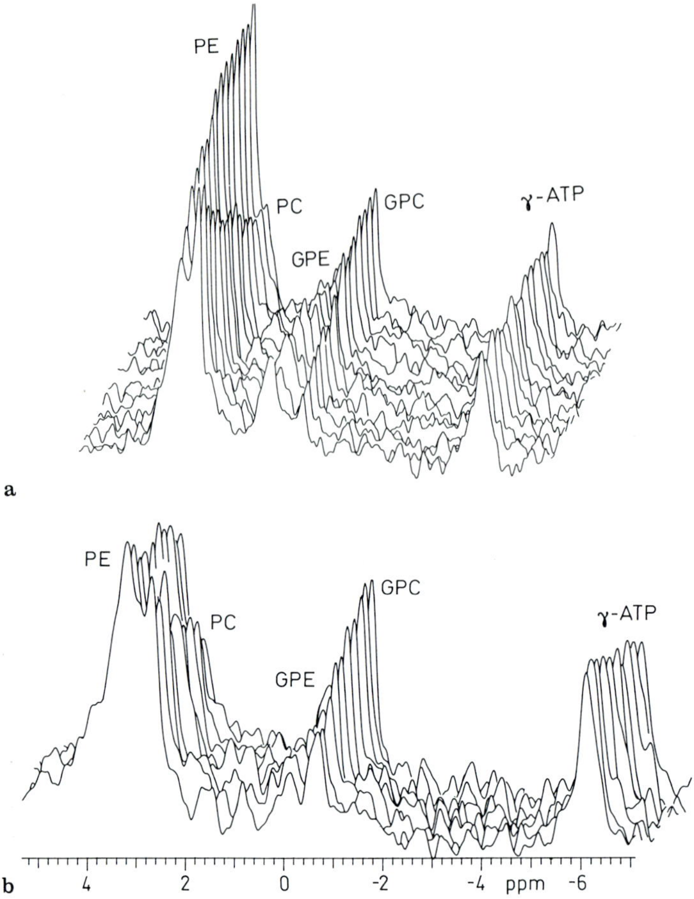


**Fig. 7. a.** Effect of ethanolamine (2 mM) on the $^{31}$P-NMR spectra at 162 MHz of perfused MDA-231 breast cancer cells at 37 °C, each spectrum represents one hour of accumulation. Note the growth of the phosphoethanolamine (PE) peak; **b.** Effect of hemicholinium-3, note the reduction of the phosphocholine (PC) peak while the PE peak is constant; also note that the $\gamma$-ATP is an internal metabolite control in each case (GPC and GPE are the corresponding glyceryl esters) [11]. Reproduced by permission

Multidrug resistance (MDR) phenomenon is one of the main reasons for treatment failure in cancer patients. Initially, substantial $^{31}$P-NMR spectral differences between the sensitive (WT) and resistant (Adr$^R$) cells were found [17]. Subsequently, $^{13}$C-NMR studies demonstrated that glycolysis is much faster in the Adr$^R$ cells at 37 °C [9]. Based on these findings drugs which might have an enhanced effects against Adr$^R$ cells were examined. The rate of phosphorylation of 2-deoxyglucose (2-DG) was faster in Adr$^R$ MCF-7 human breast cancer cells than in the WT cells, concomitant with faster ATP depletion and higher toxicity [26] (Fig. 8). Similar results were obtained in vivo, using a surface coil to detect metabolic changes following administrations of 2-DG to mice with implanted

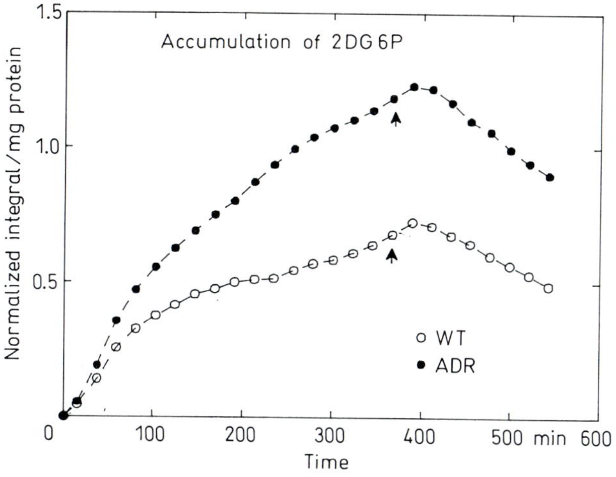


**Fig. 8.** Difference in 2DG6P accumulation in perfused WT and ADR cells by $^{31}$P-NMR spectroscopy at 162 MHz following addition of 2DG (5 mM) at time = 0. The arrow represents the replacement of 2DG-containing medium with normal medium [26]. Reproduced by permission

tumors [140]. $^{31}$P-NMR studies showed that the actions of gossypol and rhodamine-123 on these cells are different [147]. Moreover, Adr$^R$ cells exhibit cross-resistance to rhodamine, but remained sensitive to low concentrations of gossypol.

The agarose thread technique was also used in studies of the metabolism of cyclophosphamide in human histiocytic lymphoma cells [24]. Since the drug is phosphorylated, it was possible to detect the signals by $^{31}$P-NMR. Intracellular transformation of the drug derivatives to the active metabolite, phosphoramide mustard, were monitored and quantitated. ATP levels remained constant. In a recent report, data on intracellular transformation of a stable analog of the primary metabolite of cyclophosphamide (mafosfamide) in P388 mice leukemia cells, were presented [27]. It was postulated that such studies may provide information on the activity of chemotherapy in sensitive and resistant tumor cells.

$^{31}$P-NMR studies including monitoring the phosphorylation rates of 2-DG provided a plausible explanation for the effects of epidermal growth factor (EGF) on MDA-468 human breast cancer cells [18]. These cells are characterized by large number of EGF receptors, and are inhibited by EGF in vivo. EGF accelerated glucose consumption in these cells, but when they were perfused with EGF during the NMR measurements, ATP levels were stable, and they survived. Thus, the inhibitory EGF effect may be confined to cultured cells in vitro and in some cases to the natural state in vivo.

Energy metabolism and the control of substrate utilization in HeLa, and murine hybridoma and myeloma cells embedded in agarose threads were studied by Brindle and co-workers [10]. The effects of varying the contents of energy sources, glucose and glutamine, in phosphate-free perfusion solution were studied by $^{31}$P-NMR. Previously, it has been found that glucose inhibits glutamine utilization, while in cells cultured in the absence of glucose, glutamine provides more than 98% of the energy requirement. Higher intracellular Pi concentrations were measured when glutamine alone was perfused, compared to perfusing with glucose. Glucose also decreased Pi in cells already perfused with glutamine. The interpretation was that glucose regulates glutamine utilization through Pi levels, which, in turn controls glutaminase activity.

In a series of studies one group investigated the metabolism of cultured human retinal epithelial cells, and human retinoblastoma cells, in agarose threads [21, 64]. $^{31}$P-NMR spectra of retinoblastoma cells were dominated by PME signals [23], which were identified as phosphoethanolamine and phosphocholine, similar to $^{31}$P-NMR of other tumors [150, 151]. PME levels were found to correlate with feeding patterns, and it was hypothesized that they can be useful indicators of the nutritional status. However, these experiments also showed that phosphocholine depletion correlated with decrease in cellular growth, and this is in agreement with previous studies, which demonstrated that PME levels decreased as the cells approached confluency [11]. Thus it seems that an increased PME level is more an indicator of stimulated growth, than of nutritional status, and represents enhanced biosynthesis of PL in rapidly proliferating cells [152].

Glucose metabolism of retinal epithelial cells was studied by both $^{31}$P and $^{13}$C-NMR of the perfused cells, and $^{13}$C and $^{1}$H NMR of the perfusate when the cells were perfused with [1-$^{13}$C]-glucose [7]. Phosphorus spectra were used to monitor adequate perfusion, and stability of ATP levels. Simultaneously, oxygen consumption was measured. $^{13}$C-NMR data showed that lactate production proceeded at high rates, concomitant with high oxygen consumption. Moreover, high flux through the hexose monophosphate shunt was noted, which may be of significance in protection of the retina from photochemical stress. These and other experiments [18, 22, 25] nicely demonstrate the advantages of using multinuclear NMR studies of cellular metabolism, and combining NMR perfusion experiments with other modalities (oxygen consumption, glucose assays, etc.), and thus obtaining data on metabolic processes which cannot be gathered by using only one modality.

Biochemical events occurring after treating macrophages with sodium fluoride were investigated in two experimental macrophage models in agarose threads by $^{31}$P-NMR [67]. The signals which appeared in the PME region were identified by $^{31}$P and 2D $^{1}$H NMR of perchloric extracts. Fluoride perfusion caused ATP decrease, concomitant with accumulation of fructose-1-phosphate and fructose-1,6-diphosphate. These changes were explained by fluoride inhibition of glycolysis. However, the respiratory burst by $F^-$, which had been previously reported in neutrophils, could not be reproduced by these studies.

Noteworthy among $^{31}$P-NMR studies of metabolism of microorganisms in agarose threads are measurements of saturation transfer of flux between inorganic phosphate and ATP in yeast (*Saccharomyces cerevisiae*) [4]. The ratio of ATP production to oxygen consumption was close to 3, similar to the ratio found in isolated mitochondria.

Also, glucose metabolism of *Mycoplasma gallisepticum* was studied by Egan et al. [72]. These studies demonstrated the innocuous nature of embedding cells in agarose threads, as mycoplasma are very fragile, yet they maintained their integrity and metabolic stability. These results point to the regulatory role of fructose-1,6-diphosphate in mycoplasma glycolysis.

#### 2.3.2.2 Matrigel

One of the disadvantages of perfusion studies with agarose threads is the limited proliferative activity inside the threads. The gel threads technique was therefore improved with a basement membrane matrix, in which anchorage-dependent cells can multiply while being perfused [150]. Basement membrane is extracted from mice tumor, and is composed of 30% type-IV collagen, 60% laminin, 5% nidogen, 3% heparin sulfate proteoglycan, and 1% entacin [153]. Normal cells grown in basement membrane are morphologically and functionally similar to their in vivo counterpart, and cancer cells establish a "model tumor", most suitable for metabolic and pharmacologic studies (Fig. 9). Cell pellet (0.1 ml) was mixed with 2 ml of liquid basement membrane, and the mixture was extruded, as described above for agarose threads, into petri dishes. Cells were allowed to grow in the incubator until the desired densities were reached, and were then transferred to a 10-mm NMR tube. The modification of the perfusion apparatus from one which was used for the agarose threads procedure, included the insertion of large capacity filters (201) between the peristaltic pump and the tube, which ensured sterility. Thus, the cell could be perfused with fresh medium for prolonged periods (weeks), at low perfusion rate of 0.5 ml/min, and NMR spectral changes associated with proliferation were monitored [150].

The growth rate in the gel was similar to that of in vivo tumors, and concomitant elevations of phosphorus NMR signals were measured. When the cell density increased and neared confluency, the phosphomonoester (PME)/ATP ratio decreased, in agreement with previous data that PME peaks are intermediates of phospholipid synthesis, and deplete towards confluency [11]. Also, the chemical shift of the intracellular inorganic phosphate signal showed that the pH of the cells was becoming acidotic, and can be used as an indicator of tumor outgrowing.

The basement membrane technique was also used in studies of phospholipid pathways in human cancer cells [12]. Perfusion of MDA-MB-231 breast cancer cells with phospholipid precursors, ethanolamine or choline, demonstrated that the cytidylyltransferases are the rate limiting enzymes for both lecithin and cephalin synthesis in these cells. Inhibition of glycerophosphocholine phosphodiesterase by ethanolamine, allowed estimation of the flux through phospholipid degradative pathways. In MOLT-4 human lymphoma cell line, 1-$\beta$-D- arabinofuranosylcytosine perfusion resulted in cellular lysis and $^{31}$P-NMR spectra showed a significant decrease in PMEs signal. MDA-MB-231 cells are insensitive

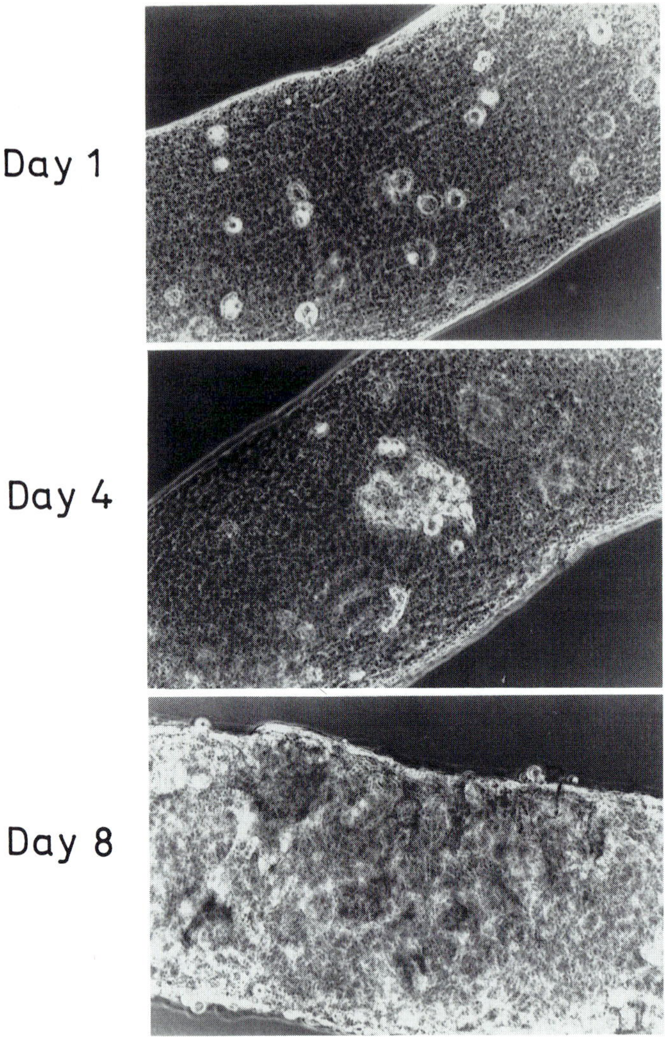


**Fig. 9.** MDA-231 breast cancer cells growing in Matrigel threads (500 $\mu$M diameter) [150]. Reproduced by permission

to the drug, and no spectral changes were noted following its addition to the perfusion solution.

Kusel et al. [154] studied the metabolic response of perfused human malignant xeroderma pigmentosum cells in basement membrane to hyperthermia. $^{31}$P-NMR data showed a reduction of total cellular phosphate together with acidification. There was a preferential decrease of high energy phosphorus compounds, ATP and phosphocreatine, while PME signals remained stable during hyperthermia.

### 2.3.2.3 Agarose and Alginate Capsules

Cells can be embedded in agarose capsules, as demonstrated by Nilsson and co-workers [155]. The cells were mixed with 5% warm liquid agarose (1:4 w/w ratio, and later changed to 1:1 w/w ratio for NMR studies [69]), and the suspension was dispersed in paraffin oil with magnetic stirring. The mixture was cooled in an ice-bath after droplets had been formed, and the agarose with the cells solidified as small capsules. These were collected after centrifugation and placed in an appropriate medium. This method was initially used to entrap microorganisms and plant cells for biotechnological applications. Subsequently, metabolism of plant cells [69, 70], and algae [156] were studied by NMR. Overall, entrapment of cells in the agarose did not alter metabolism, and they maintained the ability to proliferate. Animal cells (mouse hybridoma) were also successfully embedded in agarose capsules [157], and found to be metabolically active and to produce monoclonal antibodies. However, no NMR studies on animal or human cells in agarose capsules have been reported. It seems that with the agarose threads more cells can be put in the observed volume. Indeed, from the reported studies [70, 156] it appears that the spectra are inferior to the threads method in the S/N and the time required to achieve a quantitative spectrum, and also in magnetic field homogeneity.

In 1981 a new cell microencapsulation procedure using alginate was presented [158]. Hepatoma and pancreatic cells were suspended in sodium alginate solution, and droplets were extruded through a capillary tip into a $CaCl_2$ solution. The calcium caused immediate gelling, and the microcapsules were harvested by decantation, and enclosed in a permanent semipermeable membrane by suspension in polylysine solution. The alginate was then liquified by isotonic sodium citrate solution, and the cells were restrained in the liquid environment by the membranes. Nilsson [155] used alginate to entrap plant cells, but did not use polylysine, and the microcapsules remained in the gel phase. Similarly to agarose capsules, this technique was initially used for biotechnology purposes, and later was applied to NMR studies of perfused plant cells [69], and algae [159]. The cells maintained their metabolic stability and growth in the gel, and the procedure was thus simplified by omitting the steps of making the semipermeable membranes, and liquifying the alginate inside them.

Recently this method was adapted to NMR studies of normal and tumor human cells [15, 160, 161]. $2–3 \times 10^8$ cells were mixed with 2.5% sodium alginate (low viscosity) to yield 1.2% alginate suspension containing $1.5–2 \times 10^7$ cell/ml. The alginate was gelled to 3–4 mm diameter capsules, by using 0.1 M calcium chloride, and the exposure was kept to a minimum (up to 5 min). The perfusion apparatus was simple, consisting of a 20-mm NMR tube with the alginate capsules at the bottom, inlet and outlet tubings without an insert, a reservoir and peristaltic pump. The phosphate-free medium was preheated, saturated with 5% $CO_2$/95% air, and perfused at 2.5–3 ml/min. We have now used this method to study lymphocyte metabolism by $^{31}$P-NMR [63]. Lymphocytes are washed out of agarose threads, but can be maintained in alginate

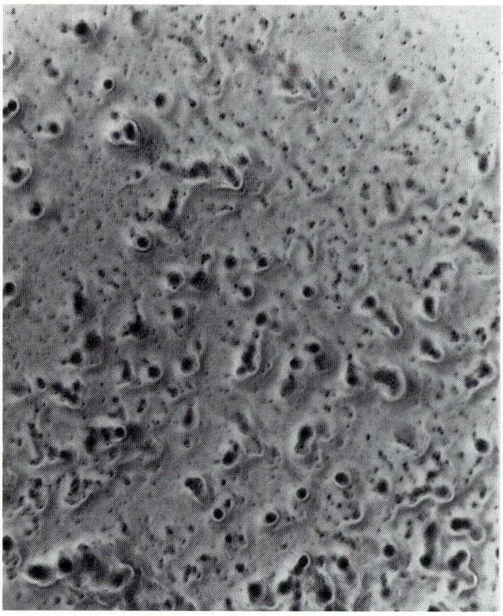

a

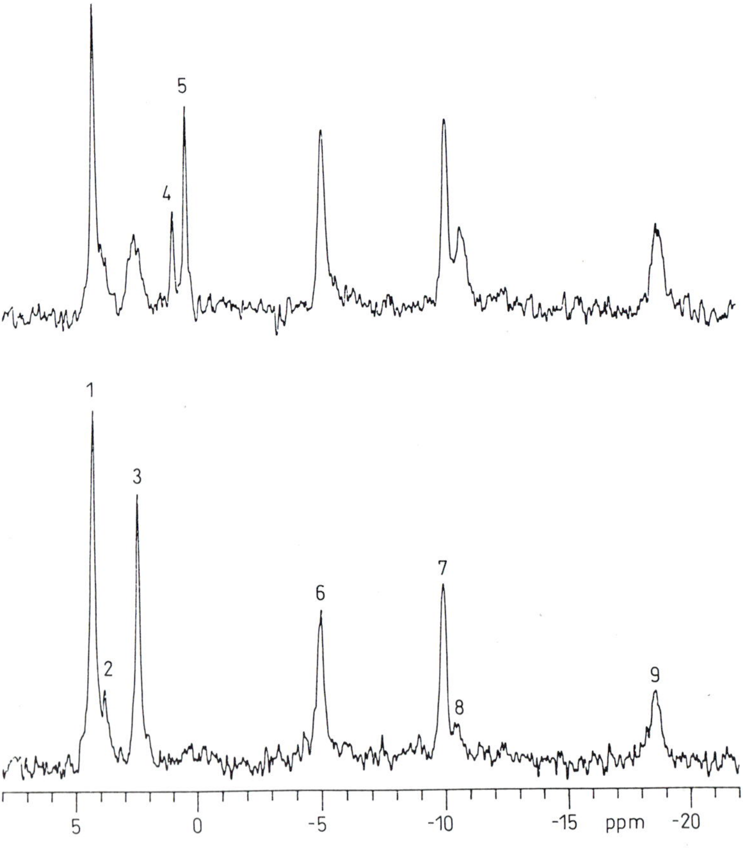


b

capsules for prolonged perfusion experiments (up to 72 h) (Fig. 10A). It was found that 2 ml of highly packed alginate capsules ($5 \times 10^8$ lymphocytes/ml capsules) can be used without metabolic compromise, at 1 ml/min perfusion rate at 37 °C. It is recommended to produce small capsules (< 1 mm diameter), in order to improve homogeneity and to increase the number of cells in the coil region. This can be achieved by using a special jet-head droplet forming apparatus, or more simply and still very efficient, by extruding the mixture through a syringe fitted with a 25G needle. Thus, a quantitative $^{31}$P spectrum could be obtained from 2 ml alginate capsules in a 10-mm tube in 10 min (Fig. 10B). However, magnetic field homogeneity is usually better with agarose threads. We have found that activation of lymphocytes, either by interleukin-2, or by phorbol esters, is associated with changes in the $^{31}$P-NMR spectra, most notably, the appearance and gradual increase of phosphodiester signals. These features were maximized in tumor infiltrating lymphocytes, cultured for 5 weeks with interleukin-2. Activation was not characterized by significant changes in ATP levels. Phospholipid pathways were also studied by perfusing alginate embedded lymphocytes with choline, and with phosphocholine cytidylyltransferase inhibitor (dapsone). $^{31}$P-NMR data indicate that choline kinase may be the rate limiting enzyme of lecithin synthesis in lymphocytes.

### 2.3.3 *Hollow Fibers and Dialysis Membranes*

The principle of these systems is separation of the cells from the flowing perfusion solution by permeable membranes, allowing diffusion of oxygen, nutrients and waste materials, but restraining the cells in the NMR probe. The first NMR hollow-fibers perfusion apparatus was developed by Gonzalez-Mendez et al. in 1982 [13]. Cells (Chinese hamster ovarian cell line) were cultured in a special culture tube, which also served as the NMR examination chamber. The culture tubes contained the hollow fibers (Vitafibers), and the cells were introduced into the interstitial space, and attached to the outer surface of the fibers. Perfusion was maintained through the fibers, and the cells had proliferated before the tube was transferred to a specially built NMR probe, with RF coils around it. High density cultures could be achieved, and prolonged experiments could be performed with this system. $^{31}$P-NMR experiments demonstrated increase in ATP when cells were perfused with fresh medium, compared to recirculating perfusion. The effects of glucose starvation, temperature changes, and variations of pH were also monitored. The technique was recently adapted for studying human melanoma cells [162], where a replaceable cartridge with the fibers was housed in a specially

◀

**Fig. 10. a**. Cut surface of alginate capsules (enlarged × 400) containing lymphocytes (small particles) with large air channels (black pores). **b**. $^{31}$P spectra of lymphocytes in alginate capsules, *lower spectrum*, normal lymphocytes and *upper spectrum* tumor-infiltrating lymphocytes (TILs) (peak assignments: *1*, phosphoethanolamine; *2*, phosphocholine; *3*, inorganic phosphate; *4*, glycerophosphoethanolamine; *5*, glycerophosphocholine; *6*, γ-ATP; *7*, α-ATP; *8*, NAD plus phosphodiester; *9*, β-ATP) [63] built probe. Growth of cells on the fibers was monitored and documented in these studies by scanning electron microscopy, demonstrating tissue-like density after 10 weeks, concomitant with increase in phosphorus metabolites detected in serial $^{31}$P-NMR measurements.

An improvement to this system was presented by Hrovat [68], in which the growth/NMR tube fitted in a conventional 15-mm vertical NMR probe. A polycarbonated tube, containing 200 fibers (200 $\mu$m outer diameter) (Fig. 11) was used, and approximately 22% of the coil volume was available for cell growth. Chicken embryo fibroblasts were maintained for over 30 days in this apparatus, and at the end they were tightly packed in tissue-like densities. It was later suggested that combining this method with closed loop NMR tubes for indirect monitoring of $^{13}$C, allows studies of carbon metabolism in intact cells without extensive design effort for probes and support systems [163].

In an attempt to increase sensitivity Karczmar developed a slightly different perfusion apparatus [19]. A bundle of cellulose acetate fibers, occupying approximately 1/6 of the sample volume, was sealed into the ends of polyethylene tubing. The bundle was placed at the bottom of 13.5 mm tube, containing a cell suspension (Chicken embryo fibroblasts and the protozoa, *Tetrahymena thermophylia*), and the perfusion solution was pumped through it at a rate of 5–10 ml/min.

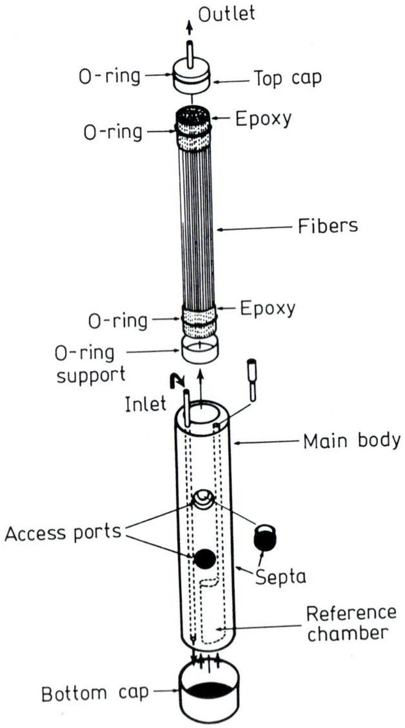


**Fig. 11.** Hollow fiber assembly for culture and perfusion of cells [68]. Reproduced by permission

A similar system was presented by Balaban et al. [164, 165]. Their apparatus had the advantage that after the cells had attached to the fibers, perfusion could be performed also through the ouside compartment, as well as through the fiber lumen. The authors reported that this design enabled studying the different sides of epithelial cells separately.

A different approach was undertaken by Boulanger et al. [65] who inserted cells (proximal tubules cells of dog kidney) into commercial dialysis fibers. The fibers were then sealed and the perfusion solution was directed over their outside. The fibers were placed in a 10-min tube, that was transferred to a special NMR probe. High flow rates were used (200 ml/min), and very rapid equilibrium of nutrients across the membranes was established. Different types and sizes of dialysis fibers were tested, and it was concluded that microfibers (200 $\mu$m internal diameter) are most suitable [66].

### *2.3.4 Spheroids*

Attachment of anchorage-dependent tumor cells to one another can produce multicellular spheroids [166, 167] (Fig. 12). Spheroids approximate many characteristics of in vivo tumors, and are potentially an efficient experimental model. Due to their size spheroids can be perfused without a restraining matrix. The first NMR study of cellular metabolism using spheroids was by Lim et al. [168]. Spheroids were formed by culturing MCF-7 human breast cancer cells, and Chinese hamster cells in agar coated culture dishes, or spinner flasks. The spinner cultures were stirred at rates of 90–100 rpm, and after the spheroids grew, they

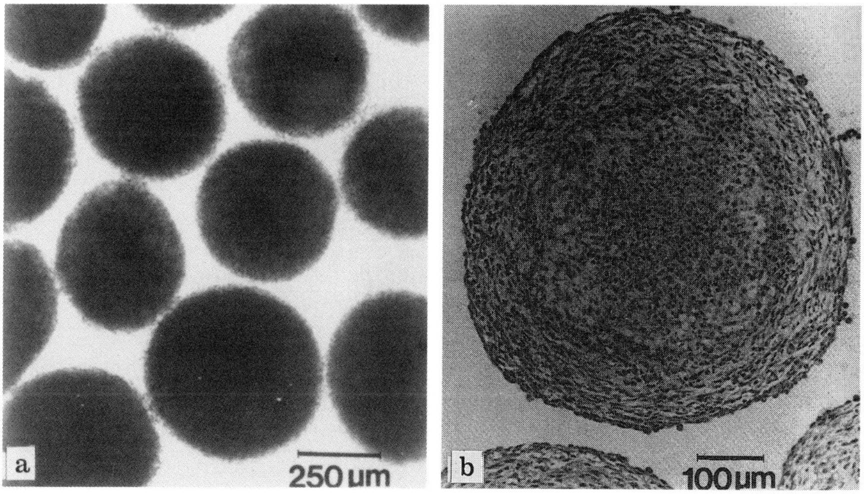


**Fig. 12.** Spheroids of cancer cells grown under controlled conditions [166]

were collected by sedimentation, and 0.5–1 ml of loosely packed spheroids were put into a 10-mm NMR tube. A simple perfusion apparatus, consisting of a rubber stopper with outflow and inflow Teflon tubings was used, and perfusion rate was 20–25 ml/min. $^{31}$P-NMR spectra obtained over a period of 80 h showed that the ATP signal increased with time. This was attributed to cell proliferation, and was regarded as evidence of the efficiency of the method. It should be pointed out however, that the S/N of the spectra presented was far from optimal. This study included also natural abundance $^{13}$C-NMR measurements of the perfused cells, although the data obtained was not informative.

An improvement in the spheroid method in order to prevent clumping of the spheroids during perfusion, they were encapsulated in agarose beads [155, 169]. 1–10 spheroids of T47D human breast cancer cells were contained in a bead (size range 200 to 800 $\mu$m), and 1.5 ml of beads (app. $10^8$ cells) were placed in a 10-mm NMR tube. Monitoring the ATP signal furnished information on cell function, and its decrease when the spheroids enlarged (to 350 $\mu$m), was used to evaluate cell necrosis. The rates of lactate production were measured by $^{13}$C-NMR when the cells were perfused with [1-$^{13}$C]-glucose, and the rates decrease as the spheroids became "older" and larger [169].

The problem of spheroid clumping was addressed differently by Halliday et al. [170] by adding a stirrer to a 20-mm NMR tube. These authors studied citrate metabolism in rat prostatic tumor cells, and also used [1-$^{13}$C]-glucose.

Differential effects of hyperthermia on 9L gliosarcoma cells attached to microcarrier beads (Cytodex) and in spheroids (200 $\mu$m) were reported by Suzuki et al. [29]. The authors claimed that $^{31}$P-NMR data showed that the critical temperature for cell survival was higher in the attached cells than in the spheroid. These results demonstrate the variability in biological activity of the various cellular preparations used in the ex vivo NMR studies of metabolism.

Another NMR study of spheroids should be mentioned; $T_1$ and $T_2$ relaxation times of $H_2O$ in $C_6$ astrocytoma cells were different at several spheroid sizes [171]. However, the spheroids were not perfused, and the metabolic stability of the cells was not established.

### 2.3.5 Other Methods

A method of anchoring cells to nonwoven fabric for NMR studies was recently presented [172]. 1.5 mm polyester fabric was surface-treated using a Corona discharge torch, cut into 1-mm squares, soaked in methanol, and covered with poly-D-lysine. These small "filters" were sterilized, and seeded with cells, which attached to them and proliferated along the fibers, forming a monolayer. 30 cm$^2$ of "filters" with cells were placed in a 10-mm NMR tube, and perfusion was carried out with a system originally used to perfuse microcarrier beads. Cells grown on the fabric could not be trypsinized, and therefore could not be directly counted. Except for the presentation of the procedure, no NMR studies of cellular metabolism using it have been reported yet, and its applicability to such experiments awaits further evaluations.

Glycolysis in high hematocrit perfused erythrocyte suspensions was investigated with a special system developed by Kuchel and co-workers [61]. This method can be regarded as a combination of suspension and perfusion. In this apparatus the whole suspension was flowing through a Plasmaflo clinical dialysis unit, which contained 1500 capillaries, with a total surface area of $0.5\,m^2$, and maximum pore size of $0.2\,\mu m$. One pump circulated the suspension from a reservoir through the center of the capillaries, while a second pump drew filtrate through the capillaries pores. The reservoir was replenished with glucose-enriched fresh buffer, gassed with 5% $CO_2$/95%$O_2$. By adjusting flow rate through the two pumps, and replenishing the perfusion solution, the hematocrit was maintained at 40%, and the total volume of the erythrocyte suspension was 80 ml.

For NMR experiments the system was modified by inserting a third circuit of tubings and pump connecting the reservoir to an NMR flow cell probe, and the suspension was flowing through it in a rate of 5 ml/min [62]. Thus, this system can be regarded as a variant of the dialysis membrane perfusion systems, which was used here for anchorage-independent cells. This system is unique because NMR data are collected from a portion of the flowing cell suspensions (made possible by the availability of large numbers of cells), while cells in all other perfusion techniques are static. $^{31}$P-NMR of perfused erythrocytes provided data on ATP and 2,3-diphosphoglycerate in fresh and stored erythrocytes under variable glucose concentrations, and showed that they can be regenerated following perfusion with inosine, pyruvate and phosphate. It was postulated that this techniques may contribute to clinical studies of the metabolic status of blood.

While it seems impossible to perfuse animal cells suspended in an NMR tube (even sophisticated filtering procedures would eventually be hampered by clumping and clogging of the filters), compressed suspensions of large plant cells were successfully perfused [71]. Sycamore cell suspension was compressed between two circular perforated Teflon plates, placed in a 25-mm NMR tube. The suspension was thus 7.5-fold concentrated to $15 \times 10^6$ cells/ml, in a 20-ml sample. The perfusion solution flowed through an inlet tubing passing through the plates to the bottom of the tube, and the effluent flowed through an outflow tubing situated above the upper plate. Plant cells could survive for 7 days while being perfused with sterile medium in this system. $^{31}$P-NMR studies were conducted on the biochemical events following sucrose deprivation of these plant cells.

### *2.3.6 Comparison of Perfusion Methods*

In this section we compare the various methods for cell perfusion that are described above. However, we do not state which is the "best" method, since preference for a particular method depends upon the cell type and the circumstances. Instead we list some of the salient features of these methods in Table 1,

**Table 1.** Comparison of perfusion techniques for NMR studies of cellular metabolism[1]

| | Suitable for anchorage-dependent cells | Suitable for anchorage-independent cells | Suitable for cellular growth studies | Metabolic stability of exptl. preparations | Suitable for studies of large molecules | Cellular density[2] (cells/ml) | S/N (general) | Field homogeneity (general) | Perfusion rates[6] (ml/min) | Technical aspects |
|---|---|---|---|---|---|---|---|---|---|---|
| A) Micro-carrier beads | yes | no | yes | [3]good | yes | $1–3 \times 10^7$ | fair | fair | 1.5–15 | simple, cheap |
| B) Gels | | | | | | | | | | |
| Agarose threads | yes | yes | no | good | [4]yes | $0.5–1 \times 10^8$ | good | good | 0.5–1 | very simple, cheap |
| Matrigel threads | yes | yes | yes | good | [5]N/D | | good | good | | simple |
| Alginate beads | yes | yes | yes | good | [4]yes | | fair | fair | | very simple, cheap |
| C) Hollow fibers and membranes | yes | [5]N/D | yes | good | no | low | fair-low | fair-low | 5–200 | complex |
| D) Spheroids | yes | no | yes | low | [5]N/D | medium | fair-low | fair-low | 10–25 | simple |

[1] For references, see appropriate section of text.
[2] Cell number in 1 ml of [cells + matrix perfusion solution]. The volume taken up by cells depends on their number, size and shape. Therefore the calculations are only approximate, and one needs to consider also, the specific cell type.
[3] Metabolic stability might deteriorate whenever cells overgrow on their attaching matrix.
[4] Established for, molecules of up to 45 kD.
[5] N/D—not determined.
[6] Depend on tube diameter.

and the advantages and disadvantages of each of the major methods are discussed below.

The *advantages* of the *bead methods* are:

(a) they are simple and easily-handled techniques, and a simple perfusion apparatus is needed;
(b) the preparations are metabolically stable for prolonged experiments;
(c) since cells actually grow on the beads they can be studied in several stages of development and thus effects on cell growth can be investigated;
(d) effects of large compounds with high molecular weight (for example proteins) can be detected; and
(e) the cells can be removed from the beads for counting, or for further experiments.

The *disadvantages* of the *bead techniques* are:

(a) the "dead" volume taken up by the beads themselves, which reduces the amount of cells in the coil region, and thus reduces S/N
(b) the less than optimal magnetic field homogeneity, most probably due to movements of the beads while being perfused;
(c) overgrowth of cell on the beads that may cause suboptimal perfusion and metabolic deterioration to the inner cell layers;
(d) the need to first let the cells attach and then grow on the beads, which is time consuming;
(e) the need to use stirring or shaking while the cells attach to the beads to prevent sedimentation that may damage the cells;
(f) perfusion rates should be low to prevent cell detachment; and
(g) most importantly, this method can be employed only with anchorage-dependent cells.

A variety of methods using gels as matrices for restraining cells for NMR studies of metabolism are currently available. None of these methods is without flaws, but an appropriate procedure can be adjusted to almost all experimental goals. The most widely used technique is with *agarose threads*; the *advantages* of this method are:

(a) it is a simple, inexpensive and quick technique;
(b) a large number of cells can be maintained in good metabolic status for prolonged experiments (24 to 36 h);
(c) the matrix occupies a relatively small volume:
(d) effects of metabolite precursors and derivatives, drugs, and physical insults on metabolism can be detected;
(e) both anchorage-dependent and -independent cells can be studied.

The *disadvantages* of the *agarose threads* are:

(a) prior to casting anchorage-dependent cells they must be harvested (usually with trypsin);

(b) the penetration of very large molecules (larger than albumin) into the threads has not been determined, and therefore it is not clear if the effects of such compounds can be studied;
(c) due to the limited growth of anchorage-dependent cells in agarose threads, this method is not optimal for studies of cellular proliferation.

The *basement membrane method* overcomes the latter obstacle, and thus, is very suitable to monitor cellular growth. The procedure is still convenient, even though more elaborate than the agarose threads, and can be utilized in any biomedical laboratory. However, since basement membrane is prepared from living tissues, it contains traces of autocrine growth factors, and it is not recommended to be used whenever one plans to investigate the effects of exogenous hormones or growth factors on cellular metabolism.

The *alginate capsule method*, which was just recently applied to human cells, seems to be a very promising procedure for NMR studies of cellular metabolism. The procedure is applicable for anchorage-dependent and -independent cells and for both small and large cells, cells multiply in the alginate, and the technique is very simple and quick. The only disadvantage so far is the less than desired field homogeneity (compared to agarose threads), probably due to the shape of the beads, and the slight movement caused by the perfusion. The role of this technique in NMR studies of metabolism remains to be seen.

The *hollow-fiber and dialysis membrane techniques* offer the significant *advantages* of enabling prolonged metabolic experiments, in which cells can be investigated by NMR during proliferation in the same container without any manipulations. Moreover, although the methods were originally developed for studying anchorage-dependent cells, theoretically they can be used for anchorage-independent cells as well. However, this has not been reported yet. As a matter of fact, these methods have not gained much popularity, and except for the publications of the various techniques and their modifications, they have not been widely applied for NMR studies of cellular metabolism. For example, an active NMR research group published a preliminary report on metabolism of HT29 human colon adenocarcinoma cells, using a dialysis membrane perfusion system [20], but in subsequent studies they used the microcarrier beads method [6, 23, 143].

Technical difficulties seem to be the main culprits; the systems are more complex than the gels and beads techniques, and require using special NMR tubes and probes. The solid support is likely to occupy a great portion of the observed volume. Indeed, from most reported spectra it is apparent that the sensitivity was low, and the length of data acquisition may limit the efficacy of monitoring metabolic processes. One of the major problems with the fibers and dialysis techniques is overgrowth of cells in "piles", clogging them and causing inadequate cell perfusion and significant back pressure. With cellulose acetate membranes (used by Karczmar and recommended by Balaban) this problem is less onerous.

Although fibers may be more durable than gels or beads during prolonged perfusions, they are far from optimal. The delicate convoluted fiber bundle used by Karczmar et al. [19] should withstand significant pressures, but may break during handling or perfusion. Also, the equilibration of nutrients through the fiber is slow, and it is permeable only to molecules of up to 1000 D, thus effects of large compounds on cell metabolism cannot be studied. Another drawback is the inability to count the number of cells (or determine protein content) of each NMR experiment, which is often required for quantitative analysis.

NMR studies of perfused *spheroids* appear to provide valuable information on "in vivo" cellular metabolism. However, the cells may not be uniformly perfused, leading to variable metabolic status and necrosis in the innermost cells. Although this may mimic an in vivo tumor heterogeneity, one should be cautious in interpreting an isolated metabolic process or a specific therapeutic effect from this experimental model.

## *2.4 Intra/Extracellular Metabolites and Transport*

A major problem encountered in NMR studies of cells and organs is discrimination between the intra- and extracellular contributions to the spectrum. The magnitude of this problem will be more pronounced when the extracellular fraction increases, and it therefore effectively depends on cell density. In ex vivo studies, this density is typically several orders of magnitude lower than for in vivo studies. For instance, in a batch of breast cancer cells embedded in agarose gel, the extracellular water volume is about a factor of 100 larger that the intracellular volume [173]. Thus a 0.01 mM constituent of the perfusion medium will be comparable in intensity to a 10-mM metabolite in the cell, when it has the same number of nuclei.

Depending on the purpose and time scale of the study, several practical approaches can be taken to avoid this problem. The most simple solution seems to be the use of medium without any solutes containing the nuclei of interest. This approach is often used for nuclei such as $^{31}$P. However, for $^{1}$H NMR, this strategy defeats the purpose of adding medium to the cells/organs in order to follow metabolic processes under physiological conditions. Also, depending on the transport rate of intracellular constituents over the plasma membrane, which may be in the order of ms, the correct intracellular metabolite ratio may be lost within a short period of time, especially when the sample is perfused. In any event, when molecular transport over the membrane is to be studied, this simple approach does not enable one to distinguish intra- and extracellular concentrations. Thus, it is preferable to design methods to separate the compartments under physiological circumstances.

Several strategies to accomplish this have been proposed, most of which are based on differences in properties of the spins of the same compounds inside and outside the cells. The properties can be chemical shift [30, 121, 174–178, 194], relaxation [179–185], and diffusion [185–187]. The latter two parameters may be interrelated when diffusion through molecular field gradients shortens T2 [188, 189]. In general, additional compounds have to be added to the medium to

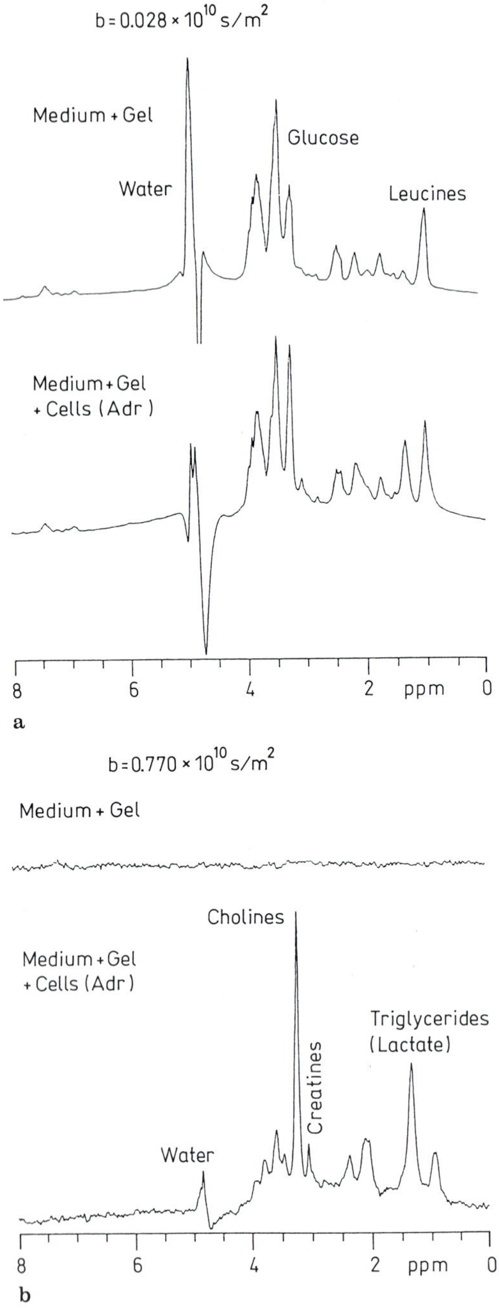


**Fig. 13.** Comparison of 200 MHz proton NMR spectra with only agarose gel threads and with gel containing cells, using different diffusion weighting factors (b). At high diffusion weighting only the intracellular metabolites are observed (Adr, multidrug-resistant MCF-7 breast cancer cells) [173]

**Table 2.** NMR studies of perfused organs (1983–1989)

| Organ | Organism | Nucleus | References |
|---|---|---|---|
| heart | rat | 31P | 199, 202, 204, 208, 210, 214, 218, 220, 221, 223, 231, 233, 234, 237, 239, 241, 242, 245, 248, 250, 254, 257, 266, 273, 276, 279, 285, 286, 287, 289, 290, 291, 294, 297, 303, 304, 305, 307, 310, 314, 320, 322, 327, 330, 332, 335, 337, 339, 340, 346, 347, 349, 352, 355, 358, 359, 364, 365, 370, 371, 373, 374 |
| heart | rat | 31P, 1H | 311 |
| heart | rat | 13C | 207, 258, 342 |
| heart | rat | 13C, 31P | 336 |
| heart | rat | 1H | 209 |
| heart | rat | 1H, 39K | 328 |
| heart | rat | 23Na, 39K | 227, 249 |
| heart | rat | 39K | 325 |
| heart | rat | 19F | 321, 366 |
| heart | rat | 19F, 5FBAPTA | 268 |
| heart | rat | 133Cs | 306 |
| heart | rat | 87Rb | 319 |
| heart | chick | 31P | 142 |
| heart | rat, guinea pig | 31P | 369 |
| heart | rabbit | 31P | 198, 205, 216, 235, 259, 271, 280, 253, 288, 317, 318 |
| heart | rabbit | 31P, 13C | 376 |
| heart | guinea pig | 31P | 197, 211, 230, 283 |
| heart | guinea pig | 13C | 228, 255, 360 |
| heart | guinea pig | 13C, 31P | 301 |
| heart | guinea pig | 1H | 299 |
| heart | ferret | 31P | 222, 302, 357, 367 |
| heart | ferret | 19F | 274, 308 |
| heart | dog | 23Na | 240 |
| heart | dog | 31P | 247, 269 |
| heart | frog | 23Na, 7Li | 263 |
| heart | frog | 23Na, 39K, 7Li | 265 |
| heart | hamster | 31P | 295, 353, 356 |
| heart | pig | 31P | 333, 345, 351 |
| kidney | rat | 31P | 201, 206, 226, 343, 362 |
| kidney | rat | 87Rb, 23Na, 31P | 338, 363 |

*(Contd.)*

**Table 2.** (*Continued*)

| Organ | Organism | Nucleus | References |
|---|---|---|---|
| kidney | rat | 23Na, 19F, 35Cl | 361 |
| kidney | rabbit | 31P | 341 |
| liver | rat | 13C | 203, 217, 225, 262, 296, 298 |
| liver | rat | 13C, 31P | 213, 244, 261, 326 |
| liver | rat | 13C, 1H | 344 |
| liver | rat | 31P | 260, 275, 281, 293, 313, 323, 329, 331, 350, 354 |
| liver | rat | 1H | 238 |
| liver | mouse | 13C | 224, 300, 375 |
| liver | mouse | 31P | 264, 277, 292, 315 |
| liver | mouse | 31P, 13C | 372 |
| liver | mouse | 1H | 229 |
| brain | rat | 31P | 212, 215, 232, 243, 368 |
| brain | rabbit | 19F | 270 |
| brain | cat | 1H, 31P | 272 |
| salivary gland | dog | 31P | 200 |
| salivary gland | rabbit | 31P | 309, 334 |
| salivary gland | rat | 39K | 348, 377 |
| mandibular gland | dog | 31P | 219 |
| mandibular gland | rat | 23Na | 278 |
| mandibular gland | rat | 31P | 282 |
| submandibular gland | dog | 31P | 236 |
| lung | pig | 31P | 246 |
| pancreas | rat | 31P | 106, 390 |
| pancreas | dog | 31P | 316 |
| bladder, uterine muscle | rabbit | 31P | 252, 324 |
| urinary bladder | rabbit | 31P | 267 |
| eye | rabbit | 13C, 31P | 253 |
| ovary | rat | 31P | 256 |
| muscle | rat | 19F | 312 |

change its properties sufficiently for the components to be distinguished. For instance, chemical shift reagents [174, 175, 177, 178], ion complexing agents with reporter groups [121, 175], and relaxation agents like $Mn^{2+}$ [179–182] have been applied. Paramagnetic shift reagents have also been used to induce magnetic field gradients around the cells, which decrease the intensity of the faster diffusion extracellular medium more than that of the intracellular [188]. The field gradients induced by oxyhaemoglobin in red cells have a similar effect [190]. An elegant method to distinguish between intra- and extra-cellular sodium ions using relaxation-allowed multiple quantum spectroscopy has been proposed by Pekar et al. [185]. Most of the above mentioned techniques have been reviewed [38, 39, 44, 121, 191] and applications determining intra- and extracellular volumes, concentrations, pH, and lifetimes have been described. However, except for the latter two instances [185, 190], the medium has to be altered by addition of sometimes highly concentrated compounds, and the physiological equilibrium may be impaired.

A technique that can be applied under physiological conditions and is applicable to any nucleus was proposed in 1976 by Andrasko [186, 187]. He used the pulsed gradient technique of Stejskal and Tanner [192, 193] to distinguish the same molecules inside and outside red cells based on their different effective diffusion constant. As first applied he used Li NMR to study lithium transport [186] and proton NMR to determine the lifetime of water spent in suspended red blood cells [187]. We have recently combined this principle with localized proton spectroscopy measurements to distinguish intracellular metabolites in perfused human breast cancer cells embedded in an agarose gel [173]. Extracellular water and small metabolites are suppressed sufficiently by diffusion weighting [195], while intracellular water is suppressed by a method outlined for in vivo studies [196] (Fig. 13). This approach allows the study of metabolism and metabolite transport under physiological conditions, as well as determination of the ratio of intra- and extracellular volumes. Thus, it should be possible to determine intracellular metabolite concentrations when a concentration standard is added to the medium and when relaxation time and diffusion corrections are applied. This principle of separating molecular components with diffusional editing will also be very useful to distinguish medium and organ signals in perfused organ studies. The ability to accurately measure only the intracellular concentration of a chosen metabolite should provide valuable metabolic information, and the ability to measure the ratio of intra/extracellular concentrations should provide insight on transport mechanisms.

## 3 Perfused Organs

The number of NMR studies of perfused organs is extremely extensive. Consequently, it is impossible to be comprehensive in this area. In order to represent a reasonable coverage of the literature, we present in Table 2 a list of such studies categorized by organ, animal, and type of study since 1983 [197–377]. Further, there have been many reviews of this general area [378–381]. The apparatus used for perfusion can vary from the simple (Fig. 14) [123] to the very complex (Fig. 15) [148].

The question that arises in relation to these studies is their relevance to the in vivo situation. Some important indications from the studies on heart, that are

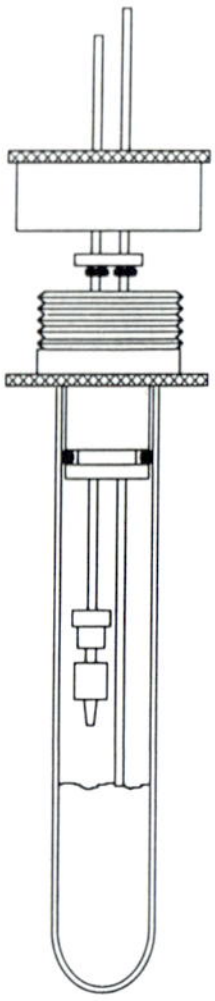

**Fig. 14.** Simple apparatus used for perfusion of ovaries in a wide bore or standard (10 mm) NMR tube [123]. Reproduced by permission

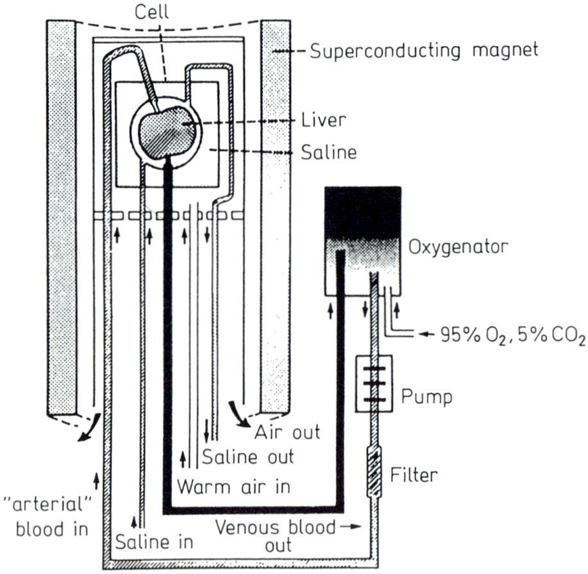


**Fig. 15.** Apparatus for perfusion of rat liver in the NMR spectrometer [148]. Reproduced by permission

the most extensive on a perfused organ, provide a useful comparison. Early studies on both Langendorff and working rat hearts gave fairly variable results [382]. Of course, it is crucial in these types of study to measure flow rate and (aortic) pressure, particularly when a triggering mechanism is used [121]. The early indication that a rhythmic variation in ATP levels could be observed [383, 384] has not been reproduced. The use of a catheter coil to monitor the internal working in in vivo heart [247, 385] gave somewhat surprising results. The ATP levels remained stable under most conditions that could be tested. For example, when the canine heart rate was increased from the basal level of 66 beat/min in stages to 182 beat/min, no significant changes in the $^{31}$P-NMR spectrum could be detected. Subsequent studies have demonstrated both the utility of the isolated perfused heart [304] and the comparison to the in vivo heart [386].

The extreme cases of major ischemia (global or local) induced by specific damage to the perfused heart may have relevance for clinical ischemia. $^{31}$P spectra showed initial reduction in PCr and concomitant increase in Pi and acidification. Subsequently, in severe cases ATP levels also decreased, and recovery of mechanical function appears to relate to the ATP level at the end of ischemia [317]. Many studies have also focussed on the protective effects of various drugs on the recovery levels of ATP under ischemic conditions [381]. Also, a number of studies have used the perfused heart model for the assessment of cardioplegic solutions. For example, the addition of the calcium antagonist nifedipine aided maintenance of ATP levels, although subsequent mechanical performance was not improved [387]. The detection of adriamycin-induced cardiotoxicity by $^{31}$P-NMR in working rat hearts showed significant decreases in PCr and ATP levels [202]. The relevance of these results for in vivo effects of these conditions and agents has not been established.

Saturation transfer methods that have been widely applied to muscle have also been extensively applied to myocardium, and have been reviewed [388]. Although the results are interesting, there is a danger in over-interpreting them, as pointed out in a careful analysis by Ugurbil and co-workers [290].

The second most studied perfused organ is perhaps the liver, usually by simple cannulation of the portal vein [121]. For liver it is necessary to perfuse with blood [121], and so far no substitute solutions have been particularly successful. Details of liver metabolism have been confirmed and expanded upon in a series of elegant studies by Sheila Cohen and associates [389] using $^{13}$C-enriched metabolite precursors. The kidney is somewhat more difficult to perfuse [116], and in general $^{31}$P-NMR results are more difficult to obtain and interpret due to the intrinsically heterogeneous nature of the kidney tissue. NMR studies of genital organs have also received some attention, notably the rat ovary and testes [256]. These results, while of interest, have shed limited information on the functioning of these organs in vivo, due mainly to the difficulty in simulating the complex endocrine factors in these organs and their subtle metabolic effects.

NMR studies of perfused rat pancreas have been used for investigating metabolic sequella of rejection reactions following allographic, as well as

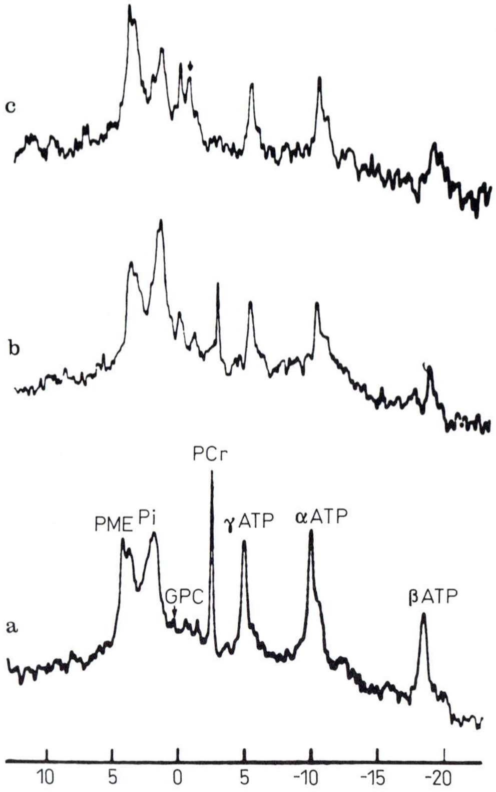


**Fig. 16 a–c.** $^{31}$P-NMR spectra at 145 MHz of perfused rat pancreas. **a)** normal gland (perfusion rate 4–6 ml/min); **b)** normal gland (perfusion rate 2–3 ml/min); **c)** experimental acute pancreatitis induced by sodium taurocholate. Arrow indicates solubilized lecithin-taurocholate complex which is typical of the disease (PME, phosphomonoester) [390]

autographic, pancreatic transplantation [106]. Experimental pancreatitis was also studied in perfused rat pancreas [390] (Fig. 16). In both studies reductions in the levels of high energy compounds correlated with the pathological process.

## 4 Conclusions

In this review we have accentuated the NMR work done on metabolism using perfused intact cells. It is our contention that this is the most appropriate control for the in vivo situation. For example, it should be noted that ATP concentrations (absolute or relative) are considered a major source of metabolic information, but these can both be elevated or reduced in vivo depending upon the circumstances [391, 392], and as indicated in Fig. 17, this can give rise to false negatives (i.e., no apparent changes).

In the context of this book our chapter focusses on the techniques that have been used to carry out studies on cells and organs, and also reviews some of the

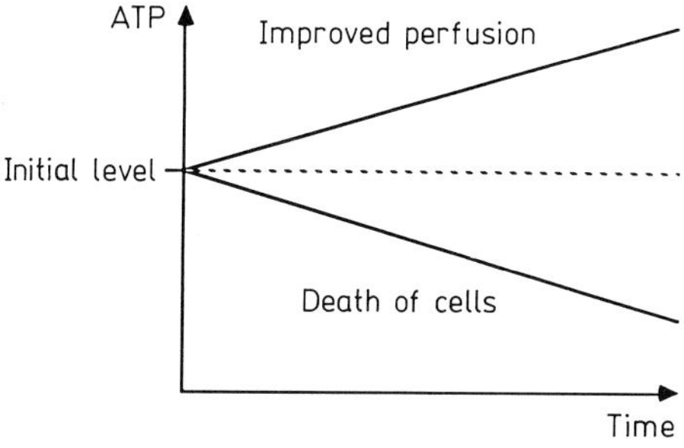


**Fig. 17.** Schematic diagram showing that since ATP levels in tumors can go either down due to cell death and/or up due to improved perfusion following cell death, then the levels could also remain constant, indicating a false negative (no response) following drug or other treatment

more important results of these studies. We do not state which methods we consider the best ones, but we hope that the reader will find the description of the methods and applications a useful source for deciding which methods to use in his own particular system. In that respect the tables that we have included may be a convenient source of comparative information.

Finally we wish to reiterate that we regard the studies on cellular metabolism as a valuable research method in themselves, divorced from considerations of clinical applications. Notably we regard this approach as a valuable means to investigate specific metabolic processes in situ, and to study the effects of growth factors, drugs, and other perturbants upon them.

## Acknowledgement

The authors wish to thank Don Crisostomo for his assistance in the preparation of this review.

## 5 References

1. Ugurbil K, Guernsey DL, Brown TR, Glynn P, Tobkes N, Edelman IS (1981) Proc Natl Acad Sci USA 78: 4843
2. Foxall DL, Cohen JS (1983) J Magn Reson 52: 346
3 Foxall DL, Cohen JS, Mitchell JB (1984) Exp Cell Res 154: 521
4 Brindle K, Krikler S (1985) Biochim Biophys Acta 847: 285
5 Painter GR, Diliberto EJ, Knoth J (1989) Proc Natl Acad Sci 86: 2239
6. Vion-Dury J, Galons JP, Fantini J, Confort-Gouny S, Kriat M, Cozzone PJ (1989) Brain Res 493: 175
7. Miceli MV, Newsome DA, Shriver GW (1990) Invest Opthamol Vis Sci 31: 277
8. Lyon RC, Faustino PJ, Cohen JS (1986) Mag Reson Med 3: 663
9. Lyon RC, Cohen JC, Faustino PJ, Megnin F, Myers CE (1988) Cancer Res 48: 870
10. Sri-Pathmannathan RM, Braddock P, Brindle KM (1990) Biochim Biophys Acta 1051: 131
11. Daly PF, Lyon RC, Faustino PJ, Cohen JS (1987) J Biol Chem 262: 14875

12. Daly PF, Zugmaier G, Sandler D, Carpen M, Myers CE, Cohen JS (1990) Cancer Res 50: 552
13. Gonzalez-Mendez R, Wemmer D, Hahn G, Wade-Jardetsky N, Jardetsky O (1982) Biochim Biophys Acta 720: 274
14. Szwergold BS, Brown TR, Freed JJ (1989) J Cell Physiol 138: 227
15. Shankar-Narayan K, Moress EA, Chatham JC, Barker PB (1990) NMR Biomed 3: 23
16. Civan MM, Williams SR, Gadian DG, Rozengurt E (1986) J Membrane Biol 94: 55
17. Cohen JS, Lyon RC, Faustino PJ, Batist G, Shoemaker M, Rubalcaba E, Cowan KH (1986) Cancer Res. 46: 4087
18. Kaplan O, Jaroszewski J, Faustino PJ, Zugmaier G, Ennis BW, Lippman MC, Cohen JS (1990) J Biol Chem 265: 13641
19. Karczmar GS, Koretsky AP, Bissell MJ, Klein MP, Weiner MW (1983) J Magn Reson 53: 123
20. Desmoulin F, Galons J-P, Canioni P, Marvaldi J, Cozzone PJ (1986) Cancer Res 46: 3768
21. Miceli MV, Kan LS, Newsome DA (1988) Biochim Biophys Acta 970: 262
22. Neeman M, Degani H (1989) Cancer Res 49: 589
23. Galons JP, Fantini J, Vion-Dury J, Cozzone PJ, Canioni P (1990) Int J Cancer 45: 168
24. Boyd VL, Robbins JD, Egan W, Ludeman SM (1986) J Med Chem 29: 1206
25. Neeman M, Degani H (1989) Proc Natl Acad Sci USA 55: 85
26. Kaplan O, Navon G, Lyon RC, Faustino PJ, Straka EJ, Cohen JS (1990) Cancer Res 50: 544
27. Sonawat HM, Leibfritz D, Engel J, Hilgard P (1990) Biochim Biophys Acta 1052: 36
28. Melner MH, Sawyer ST, Evanochko WT, Ng, TC, Glickson JD, Puett D (1983) Biochemistry 22: 2039
29. Suzuki N, Kwee IL, Nakada T (1989) Proc 8th Ann Mtg of SMRM, Amsterdam, Aug 1989, vol 1, p 417
30. Brown FF, Campbell ID, Kuchel PW, Rabenstein DL (1077) FEBS Lett 82: 12
31. Brown FF, Campbell ID (1980) Philos Trans R Soc Lond Ser B 289: 395
32. Campbell ID, Dobson CM, Jeminet G, Williams RJP (1974) FEBS Lett 49: 115
33. Isab AA, Rabenstein DL (1979) FEBS Lett 106: 325
34. Daniels A, Williams RJP, Wright PE (1976) Nature 261: 321
35. Rabenstein DL, Isab AA (1979) J Magn Reson 36: 281
36. Rabenstein DL, Isab AA, Brown DW (1980) J Magn Reson 41: 361
37. Rabenstein DL, Millis KK, Strauss EJ (1988) Anal Chem 60: 1380A
38. Rabenstein DL (1984) J Biochem Biophys Methods 9: 277
39. Kuchel PW (1989) In: Field LD, Sternhel S (eds) Analytical NMR Wiley, London, p 157
40. Hore PJ (1983) J Magn Reson 55: 283
41. van Zijl PCM, Moonen CWT (1990) other chapter in this book
42. Kuchel PW (1981) CRC Crit Rev Anal Chem 12: 155
43. Brindle KM, Campbell ID (1987) Q Rev Biophys 19: 159
44. Fabry ME (1987) In: Gupta RK (ed) NMR spectroscopy of cells and organisms. CRC Press, Boca Raton, Vol 1, p 69
45. Agris PF, Campbell ID (1982) Science 216: 1325
46. Navon G, Burrows H, Cohen JS (1983) FEBS Lett 162: 320
47. Irving MG, Simpson SJ, Brooks RS, Holmes RS, Doddrell DM (1985) Int J Biochem 17: 471
48. Reglinksi J, Smith WE, Suckling CJ, Al-Kabban M, Watson ID, Stewart MJ (1987) FEBS Lett 214: 351
49. Reglinksi J, Smith WE, Suckling CJ, Al-Kabban M, Stewart MJ, Watson ID (1988) Clin Chim Acta 175: 285
50. Mountford CE, Wright LC (1988) Trends Biochem Sci 13: 172
51. Mountford CE, Tattersall MHN (1987) Cancer Surv 6: 285
52. Bloom M, Holmes KT, Mountford CE, Williams PG (1986) J Magn Reson 69: 73
53. Holmes KT, Williams PG, King NJC, May GL, Dyne M, Bloom M, Mountford CE (1987) Magn Reson Med 4: 567
54. May GL, Wright LC, Holmes KT, Williams PG, Smith ICP, Wright PE, Fox RM, Mountford CE (1986) J Biol Chem 261: 3048
55. Williams PG, Helmer MA, Wright LC, Dyne RM, Fox KT, Holmes GL, May GL, Mountford CE (1985) FEBS Lett 192: 159
56. Wright LC, May GL, Dyne M, Mountford CE (1986) FEBS Lett 203: 164
57. Mountford CE, May GL, Wright LC, MacKinnon WB, Dyne M, Holmes KT, van Haaften-Day C, Tattersall MHN (1987) Lancet 1: 829
58. Wieczorek AJ, Rhyner G, Block LH (1985) Proc Natl Acad Sci USA 82: 3455

59. Skipski VP, Barclay M, Archibald FM, Stock CC (1971) Biochem Pharmacol 10: 112
60. Poste GW, Nicolson GL (1980) Proc Natl Acad Sci USA 77: 399
61. Kuchel PW, Chapman BE, Lovric VA, Raftos JE, Stewart IM, Thorburn DR (1984) Biochim Biophys Acta 805: 191
62. Chapman BE, Kuchel PW, Lovric VA, Raftos JE, Stewart IM (1985) Br J Hematol 61: 385
63. Kaplan O, Cohen JS (1991) J Biol Chem 265: 20712
64. Miceli MV, Kan L, Newsome DA (1987) Invest Opthamol Vis Sci 28 (supp): 26
65. Boulanger Y, Vinay P, Phan Viet MT, Guardo R, Desroches M (1985) Magn Reson Med 2: 495
66. Boulanger Y, Vinay P (1986) Cell Biol 64: J869
67. Ducrocq C, Lenfant M, Werner GH, Gillet B, Beloeil JC (1987) Biochem Biophys Res Comm 147: 519
68. Hrovat MI, Wade CG, Hawkes SPA (1985) J Magn Reson 61: 409
69. Vogel HJ, Brodelius P (1984) J Biotechnol 1: 159
70. Brodelius P, Vogel HJ (1985) J Biol Chem 260: 3556
71. Roby C, Martin J-B, Bligny R, Douce R (1987) J Biol Chem 262: 5000
72. Egan W, Barile M, Rottem S (1986) FEBS Lett 204: 373
73. Navon G, Ogawa S, Shulman R, Yamane T (1977) Proc Natl Acad Sci USA 74: 87
74. Burt CT, Moore RR, Roberts MF (1983) J Magn Reson 53: 163
75. Glonek T, Kopp SJ, Kot E, Pettigrew JW, Harrison WH, Cohen MM (1982) J Neurochem 39: 1210
76. Graham RA, Meyer RA, Szwergold BS, Brown TR (1987) J Biol Chem 262: 35
77. Evanochko WT, Sakai TT, Ng TC, Krishna NR, Kim HD, Zeidler RB, Ghanta VK, Brockman RW, Schiffer LM, Braunschweiger PG, Glickson JD (1984) Biochim Biophys Acta 805: 104
78. Kushnir T, Kaplan O, Askenasy N, Navon G (1989) Mag Reson Med 10: 119
79. Askenazy N, Kushnir T, Kaplan O, Navon G (1990) NMR Biomed, (in press)
80. Lowry OH, Passonneau JV (1972) A flexible system of enzymatic analysis. Academic Press, New York, p 120
81. Barny M, Glonek T (1982) In: Fredericksen DW, Cunningham LW (eds) Methods Enzymol 85: 624
82. Bass MB, Fromm HJ (1985) Anal Biochem 45: 292
83. Veech RL, Harris RL, Veloso D, Veech EH (1973) J Neurochem 20: 183
84. Cerdan S, Parilla R, Santoro J, Rico M (1985) FEBS Lett 187: 167
85. Van Waarde A, Stromski ME, Thulin G, Gaudio KM, Kashgarin M, Shulman RG (1989) Am J Physiol 256: F298
86. den Hollander JR, Ugurbil K, Shulman RG (1986) Biochemistry 25: 212
87. Hoult DI, Busby SJW, Gadian DG, Radda GK, Richards RE, Seely PJ (1974) Nature 252: 285
88. Behar KL, den Hollander HA, Stromski ME, Ogino T, Shulman RG, Petroff OAC, Prichard JW (1983) Proc Natl Acad Sci USA 80: 4945
89. Fan TWM, Higashi RM, Lane AN, Jardetsky O (1986) Biochim Biophys Acta 882: 154
90. Petroff OAC, Ogino T, Alger JR (1988) J Neurochem 51: 163
91. Petroff OAC, Spencer DD, Alger JR, Prichard W (91989) Neurology 39: 1197
92. Ogawa S, Lee TM, Glynn P (1986) Arch Biochem Biophys 248: 43
93. Pettigrew JW, Kopp SJ, Minshew NJ, Glonek T, Feliksik JM, Tow JP, Cohen MM (1987) J Neuropathol Exp Neurol 46: 419
94. Nakada T, Kwee IL, Suzuki NI, Houkin K (1989) Mag Reson Med 12: 172
95. Behar KL, den Hollander JA, Petroff OAC, Hetherington HP, Prichard JW, Shulman RG (1985) J Neurochem 44: 1045
96. Peeling J, Wong D, Sutherland GR (1989) Stroke 20: 633
97. Ross BD, Higgins RJ, Boggan JE, Knittel B, Garwood M (1988) Magn Reson Med 6: 403
98. Cohen SM (1983) J Biol Chem 258: 14294
99. Cohen SM (1987) Biochemistry 26: 563
100. Cerdan S, Kunnecke B, Dolle A, Seelig J (1988) J Biol Chem 263: 1664
101. Ling MF, Brauer M (1990) Biochim Biophys Acta 1051: 151
102. Kopp SJ (1983) Toxicol Appl Pharmacol 70: 273
103. Mally Cr, Sherry AD, Jeffrey FM (1988) J Biol Chem 263: 6964
104. Jans AW, Grunewald RW, Kinne RK (1989) Mag Reson Med 9: 419
105. Kaplan O, Kushnir T, Sandbank U, Navon G (1987) J Sug Res 43: 172
106. Morris GE, Williams SR, Proctor E, Gadian DG, Browse NL (1989) Transplantation 47: 779
107. Venkatasubramanian PN, Arus C, Barany M (1986) Clin Physiol Biochem 4: 285
108. Greiner JV, Braude LS, Glonek T (1985) Exp Eye Res 40: 335

109. Apte DW, Koutalos Y, McFarlane DK, Dawson MJ, Ebrey TG (1989) Biophys J 56: 447
110. Navon G, Gogol E, Weissenberg R (1985) Arch Androl 15: 153
111. Merchant TE, Gierke LW, Meneses P, Glonek T (1988) Cancer Res 48: 5112
112. Navon G, Navon R, Shulman RG, Yamane T (1978) Proc Natl Acad Sci USA 75: 891
113. Evans FE, Kaplan O (1978) Proc Natl Acad Sci USA 74: 4909
114. Corbett RJT, Nunnally RL, Giovanella BC, Antich PP (1987) Cancer Res 47: 5065
115. Galons JP, Fantini J, Vion-Dury J, Cozzone PJ, Canioni P (1989) Biochimie 71: 949
116. Kalderon B, Gopher A, Lapidot A (1986) FEBS Lett 204: 29
117. Kaplan O, van Zijl PCM, Cohen JS (1990) Biochem Biophys Res Commun 169: 383
118. Bines SD, Tomasovic SP, Frazer JW, Boddie AW (1985) J Surg Res 38: 546
119. Folch J, Ascoli L, Less M, Meath JA, Lebaron FN (1951) J Biol Chem 191: 833
120. Shulman RG, Brown TR, Ugurbil K, Ogawa S, Cohen SM, den Hollander JA (1979) Science 205: 160
121. Morris PG (1988) In: Webb GA (ed) Ann Rep on NMR spectroscopy, vol 20, Academic Press, New York, p 1
122. Boddie AW, Frazer JW, Tomasovic SP, Dennis L (1989) J Surg Res 46: 90
123. Cohen JS, Lyon RC, Daly PF (1988) Methods Enzymol. 48: 435
124. Ugurbil K, Rottenberg H, Glynn P, Shulman RG (1982) Biochemistry 21: 1068
125. Balaban RS, Gadian DG, Radda GK, Wong GG (1981) Anal Biochem 116: 450
126. Jentoft JE, Town CD (1985) J Cell Biol 101: 778
127. Santos H, Turner DL (1986) J Mag Reson 68: 354
128. Labotka RJ, Warth JA, Winecki V, Omachi A (1985) Anal Biochem 147: 75
129. Matwiyoff NA, Needham TE (1972) Biochem Biophys Res Commun 49: 1158
130. Eakin RT, Morgan LO, Gregg CT, Matwiyoff NA (1972) FEBS Lett 28: 259
131. Moon RB, Richards JH (1973) J Biol Chem 248: 7276
132. Henderson TO, Costello AJR, Omaci A (1974) Proc Natl Acad Sci USA 71: 2487
133. Ugurbil K, Brown TR, den Hollander JA, Glynn P, Shulman RG (1978) Proc Natl Acad Sci USA 75: 3742
134. den Hollander JA, Brown TR, Ugurbil K, Shulman RG (1979) Proc Natl Acad Sci USA 76: 6096
135. Cohen SM, Shulman RG (1980) Phil Trans R Soc Lond Ser B 289: 407
136. Gupta RK, Gupta P, Moore RD (1984) Ann Rev Biophys Bioeng 13: 221
137. Brophy PJ, Hayer KM, Riddell FG (1983) Biochem J 210: 961
138. Malet-Martino MC, Lame F, Vialaneix JP, Palevedy C, Hollande E, Martino R (1986) Cancer Chemother 18: 5
139. Legerton TL, Kanamori K, Weiss RL, Roberts JD (1983) Biochemistry 22: 899
140. Navon G, Lyon RC, Kaplan O, Cohen JS (1989) FEBS Lett 247: 86
141. Cohen JS (1985) In: Govil T, Khetrapal CL, Saran A (eds) Magnetic resonance in biology and medicine. Tata-McGraw Hill, New Delhi, India, p 77
142. Hochhauser E, Barak J, Kushnir T, Navon G, Meyer MS, Edelstein S, Ben Bassat MB, Vidne BA (1989) Angiology 40: 300
143. Fantini J, Galons JP, Marvaldi J, Cozzone PJ, Canioni P (1987) Int J Cancer 39: 255
144. Glynn P, Lee TM, Ogawa S (1987) Ann NY Acad Sci 508: 474
145. Egan WM (1987) In: Gupta RK (ed) NMR spectroscopy of cells and organisms. CRC Press, Boca Raton, vol 1: 99
146. Kaplan O, Jaroszewski J, Clarke R, Fairchild CR, Schoenlien P, Goldenberg S, Gottesman MM, Cohen JS (1991) Cancer Res 51: 1638
147. Jaroszewski J, Kaplan O, Cohen JS (1990) Cancer Res 50: 6936
148. Iles RA, Griffiths JR, Stevens AN, Gadian DG, Porteous RR (1980) Biochem J 192: 191
149. Geoffrion Y, Lareau S, Deslaurier R, Butler K, Pass M, Smith ICPA (1985) Mag Reson Med 2: 65
150. Daly PF, Lyon RC, Straka EJ, Cohen JS (1988) FASEB J 2: 2596
151. Evelhoch JL, Keller NA, Corbett TH (1987) Cancer Res 47: 3396
152. Daly P, Cohen JS (1989) Cancer Res 49: 770
153. Kleinman HK, McGarvey ML, Hassell JR, Star VL, Cannon FB, Laurie GW (1986) Biochemistry 25: 312
154. Kusel AC, Grascew G, Thielmann H, Lorenz W, Hull WE (1989) Proc 8th Ann of SMRM Meeting, Amsterdam
155. Nilsson K, Birnbaum S, Flygare S, Linse L, Schroder U, Jeppsson U, Larsson PO, Mosbach K, Brodelius P (1983) J Appl Microbiol Biotech 17: 319
156. Bental M, Pick U, Avron M, Degani H (1990) Eur J Biochem 188: 111

157. Nilsson K, Scheirer W, Merten OW, Ostberg L, Liehl E, Katinger HWD, Moshbach K (1983) Nature 302: 629
158. Lim F, Moss RD (1981) J Pharm Sci 70: 351
159. Oren-Shamir M, Avron M, Degani H (1988) FEBS Lett 233: J214
160. Barker PB, Shanker-Narayan K, Chatham JC, Moress EA (1988) 7th Ann Meeting of SMRM, San Francisco
161. Ross B. Derby KA, Tropp J, Hawrysko C, Yamagata H, Shankar-Narayan K, Jacques DB, Ingram M (1989) 8th Ann Meeting of SMRM, Amsterdam
162. Minichello MM, Albert DM, Kolodny NH, Lee MS, Craft JJ (1989) Mag Reson Med 10: 96
163. O'Leary DJ, Hawkes SP, Wade CG (1987) Mag Reson Med 5: 572
164. Knazek RA, Gulliano PM, Kohler PO, Dedrick RI (1972) Science 178: 65
165. Balaban RS (1984) Am J Physiol 246: C10
166. Deen DF, Hoshino T, Williams ME, Muraoka I, Knebel KD, Barker M (1980) JNCI 64: 1373
167. Freyer JP (1988) Cancer Res 48: 2432
168. Lim P-J, Blumenstein M, Mikkelsen RB (1987) J Magn Res 73: 399
169. Ronen S, Degani H (1989) Mag Reson Med 12: 274
170. Halliday KR, Clinard EH, Huston YY, Freyer JP, Fink NH, Schor PL, Fenoglio-Presiser C, Sillerud LO (1989) Proc 8th Ann Meeting of SMRM, Amsterdam
171. Megyesi JF, Del Maestro RF (1988) Biochem Cell Biol 66: 1100–1109
172. Neeman M, Rushkin E, Kadouri A, Degani H (1988) Mag Reson Med 7: 236–242
173. van Zijl PCM, Moonen CWT, Faustino PJ, Kaplan O, Pekar J, Cohen JS (1990) Proc Natl Acad Sci USA 88: 3228
174. Cowan BE, Sze DY, Mai MT, Jardetsky O (1985) FEBS Lett 184: 130
175. Kirk K, Kuchel PW (1986) Stud Biophys 116: 139
176. Smith GA, Hesketh TR, Metcalfe JC, Feeney J, Morris PG (1983) Proc Natl Acad Sci USA 80: 7178
177. Gupta RK, Gupta RK (1982) J Magn Reson 47: 344
178. Pike MM, Simon SR, Balschi JA, Springer CS (1982) Proc Natl Acad Sci USA 79: 810
179. Conlen T, Outhred R (1972) Biochim Biophys Acta 288: 354
180. Fabry ME, Eisenstadt M (1975) Biophys J 15: 101
181. Pirkle JL, Ashley DL, Goldstein JH (1979) Biophys J 25: 389
182. Benga G, Pop VI, Popescu O, Ionescu M, Mihele V (1983) J Membr Biol 76: 129
183. Eisenstadt M (1980) J Magn Reson 39: 275
184. Rayson BM, Gupta RK (1985) J Biol Chem 260: 7276
185. Pekar J, Renshaw PF, Leigh JS (1987) J Magn Reson 72: 159
186. Andrasko J (1976) J Magn Reson 21: 479
187. Andrasko J (1976) Biochim Biophys Acta 428: 428
188. Brindle KM, Brown FF, Campbell ID, Grathwohl C, Kuchel PW (1979) Biochem J 37: 180
189. Packer KJ (1973) J Magn Reson 9: 438
190. Fabry ME, San George RC (1983) Biochemistry 22: 4119
191. Jacobsen L, Cohen JS (1982) In: JS Cohen (ed) Noninvasive probes of tissue metabolism. Wiley-Interscience, New York, p 5
192. Stejskal EO, Tanner JE (1965) J Chem Phys 42: 288
193. Tanner JE, Stejskal J (1968) J Chem Phys 49: 1768
194. Robenstein DL, Isab AA (1982) Anal Biochem 121: 423
195. Moonen CTW, van Zijl PCM (1990) J Magn Reson 87: 18
196. Moonen CTW, van Zijl PCM (1990) J Magn Reson 88: 28
197. Nakazawa M, Katano Y, Imai S, Matsushita K, Ohuchi M (1982) Cardiovasc Pharmacol J 4: 700
198. Jacobus WE, Pores IH, Lucas SK, Weisfeldt ML, Flaherty JT (1982) J Mol Cell Cardiol 14 Suppl 3: 13
199. Matthews PM, Bland JL, Gadian DG, Radda GK (1982) Biochim Biophys Acta 721: 312
200. Murakami M, Imai Y, Seo Y, Morimoto T, Shiga K, Watari H (1983) Biochim Biophys Acta 762: 19
201. Freeman D, Bartlett S, Radda G, Ross B (1983) Biochim Biophys Acta 762: 325
202. Ng TC, Daugherty JP, Evanochko WT, Digerness SB, Durant JR, Glickson JD (1983) Biochem Biophys Res Commun 110: 339
203. Sillerud LO, Shulman RG (1983) Biochemistry 22: 1087
204. Seymour AM, Bailey IA, Radda GK (1983) Biochim Biophys Acta 762: 525

205. Whitman G, Kieval R, Wetstein L, Seeholzer S, McDonald G, Harken A (1983) J Surg Res 35: 332
206. Rhodes RS, Jentoft JE, Barr RG, Robinson AV (1983) J Surg Res 35: 373
207. Chance EM, Seeholzer SH, Kobayashi K, Williamson JR (1983) J Biol Chem 258: 13785
208. Dhasmana JP, Digerness SB, Geckle JM, Ng TC, Glickson JD, Blackstone EH (1983) J Cardiovasc Pharmacol 5: 1040
209. Ugurbil K, Petein M, Maidan R, Michurski S, Cohn JN, From AH (1984) FEBS Lett 167: 73
210. Pieper GM, Murray WJ, Salhany JM, Wu ST, Eliot RS (1984) Biochim Biophys Acta 803: 241
211. Brooks WM, Willis RJ (1983) J Mol Cell Cardiol 15: 495
212. Ackerman JJ, Berkowitz BA, Deuel RK (1984) Biochem Biophys Res Commun 119: 913
213. Cohen SM (1984) Fed Proc 43: 2657
214. Lavanchy N, Martin J, Rossi A (1984) Cardiovasc Res 18: 573
215. Kopp SJ, Krieglstein J, Freidank A, Rachman A, Seibert A, Cohen MM (1984) J Neurochem 43: 1716
216. Renlund DG, Gerstenblith G, Lakatta EG, Jacobus WE, Kallman CH, Weisfeldt ML (1984) J Mol Cell Cardiol 16: 795
217. Albert K, Kruppa G, Zeller KP, Bayer E, Hartmann F (1984) Z Naturforsch 39: 859
218. Lavanchy N, Martin J, Rossi A (1984) FEBS Lett 178: 34
219. Murakami M, Seo Y, Nakahari T, Mori H, Imai Y, Watari H (1984) Jpn J Physiol 34: 587
220. Kupriyanov VV, Ya Steinschneider A, Ruuge EK, Kapel'ko VI, Yu Zueva M, Lakomkin VL, Smirnov VN, Saks VA (1984) Biochim Biophys Acta 805: 319
221. Garlick PB, Brown TR, Sullivan RH, Ugurbil K (1983) J Mol Cell Cardiol 15: 255
222. Morris PG, Allen DG, Orchard CH (1985) Adv Myocardiol 5: 27
223. Bittl JA, Ingwall JS (1985) J Biol Chem 260: 3512
224. Pass MA, Geoffrion Y, Deslauriers R, Butler KW, Smith IC (1984) J Biochem Biophys Methods 10: 135
225. Desmoulin F, Canioni P, Cozzone PJ (1985) FEBS Lett 185: 29
226. Manganaro M, Miccheli A, Delfini M, Gaudio E, Marinozzi G, Conti F (1985) Cell Mol Biol 31: 89
227. Pike MM, Frazer JC, Dedrick DF, Ingwall JS, Allen PD, Springer CS Jr, Smith TW (1985) Biophys J 159
228. Sherry AD, Nunnally RL, Peshock RM (1985) J Biol Chem 260
229. Jue T, Arias-Mendoza F, Gonnella NC, Shulman GI, Shulman RG (1985) Proc Natl Acad Sci USA 82: 5246
230. Brooks WM, Willis RJ (1985) J Mol Cell Cardiol 17: 747
231. Pieper GM, Wu ST, Salhany JM (1985) J Mol Cell Cardiol 17: 775
232. Litt L, Gonzalez-Mendez R, Severinghaus JW, Hamilton WK, Shuleshko J, Murphy-Boesch J, James TL (1985) J Cereb Blood Flow Metab 5: 537
233. Nakazawa M, Tamatsu H, Tsuchihashi H, Nagatomo T, Imai S (1985) Jpn J Pharmacol 39: 51
234. Degani H, Laughlin M, Campbell S, Shulman RG (1985) Biochemistry 24: 5510
235. Gard JK, Kichura GM, Ackerman JJ, Eisenberg JD, Billadello JJ, Sobel BE, Gross RW (1985) Biophys J 48: 803
236. Nakahari T, Seo Y, Murakami M, Mori H, Miyamoto S, Imai Y, Watari H (1985) Jpn J Physiol 35: 729
237. Prentice RC, Kopp SJ (1985) Toxicol Appl Pharmacol 81: 491
238. Allars H, Coleman MD, Norton RS (1985) Eur J Drug Metab Pharmacokinet 10: 253
239. Kopp SJ, Daar AA, Prentice RC, Tow JP, Feliksik JM (1986) Toxicol Appl Pharmacol 82: 200
240. Cannon PJ, Maudsley AA, Hilal SK, Simon HE, Cassidy F (1986) J Am Coll Cardiol 7: 573
241. Koretsky AP, Wang S, Klein MP, James TL, Weiner MW (1986) Biochemistry 25: 77
242. Kingsley-Hickman P, Sako EY, Andreone PA, St. Cyr JA, Michurski S, Foker JE, From AH, Petein M, Ugurbil K (1986) FEBS Lett 198: 159
243. Kariman K, Chance B, Burkhart DS, Bolinger LA (1986) Brain Res 366: 300
244. Canioni P, Desmoulin F, Galons JP, Bernard M, Fontanarava E, Cozzone PJ (1985) Arch Int Physiol Biochim 93: 119
245. Bernard M, Menasche P, Canioni P, Fontanarava E, Geyer RP, Piwnica A, Cozzone P (1986) Arch Int Physiol Biochim 93: 97
246. Pillai RP, Buescher PC, Pearse DB, Sylvester JT, Eichhorn GL (1986) Magn Reson Med 3: 467
247. Kantor HL, Briggs RW, Metz KR, Balaban RS (1986) Am J Physiol 251: H171
248. Lavanchy N, Martin J, Giacomelli M, Rossi A (1986) Eur J Pharmacol 125: 341
249. Fossel ET, Hoefeler H (1986) Magn Reson Med 3: 534

250. From AH, Petein MA, Michurski SP, Zimmer SD, Ugurbil K (1986) FEBS Lett 206: 257
251. Komabayashi T, Watanabe K, Izawa T, Tsuboi M (1986) Res Commun Chem Pathol Pharmacol 53: 249
252. Kushmerick MJ, Dillon PF, Meyer RA, Brown TR, Krisanda JM, Sweeney HL (1986) J Biol Chem 261: 14420
253. Willis JA, Schleich T (1986) Exp Eye Res 43: 329
254. Lavanchy N, Martin J, Rossi A (1986) J Mol Cell Cardiol 18: 931
255. Brainard JR, Hoekenga De, Hutson JY (1986) Magn Reson Med 3: 673
256. Haseltine FP, Arias-Mendoza F, Kaye AM, Degani H (1986) Magn Reson Med 3: 796
257. Lavanchy N, Martin J, Rossi A (1986) Presse Med 15: 1758
258. Malloy CR, Sherry AD, Jeffrey FM (1987) FEBS Lett 212: 58
259. Freeman D, Mayr H, Schmidt P, Roberts JD, Bing RJ (1987) Biochim Biophys Acta 927: 350
260. Desmoulin F, Cozzone PJ, Canioni P (1987) Eur J Biochem 162: 151
261. Cohen SM (1987) Biochemistry 26: 573
262. Cohen SM (1987) Biochemistry 26: 581
263. Burstein D, Fossel ET (1987) Magn Reson Med 4: 261
264. Lanir A, Clouse ME, Lee RG (1987) Transplantation 43: 786
265. Burstein D, Fossel ET (1987) Am J Physiol 252: H1138
266. Lavanchy N, Martin J, Rossi A (1987) Arch Int Pharmacodyn Ther 286: 97
267. Fisher MJ, Dillon PF (1987) Circ Res 60: 472
268. Steenbergen C, Murphy E, Levy L, London RE (1987) Circ Res 60: 700
269. Zweier JL, Jacobus WE (1987) J Biol Chem 262: 8015
270. Mills P, Sessler DI, Moseley M, Chew W, Pereira B, James TL, Litt L (1987) Anesthesiology 67: 169
271. McAuliffe JJ, Hickey PR (1987) Anesthesiology 67: 231
272. Sutton LN, McLaughlin AC, Kemp W, Schnall MD, Cho BK, Langfitt TW, Chance B (1987) J Neurosurg 67: 381
273. Carlier PG, Jacobstein MD, Portman MA, Fouad F, Tarazi RC, Gerken TA (1987) J Hypertens Suppl 4: S95
274. Marban E, Kitakaze M, Kusuoka H, Porterfield JK, Yue DT, Chacko VP (1987) Proc Natl Acad Sci USA 84: 6005
275. Thoma WJ, Ugurbil K (1987) Biochim Biophys Acta 893: 225
276. Katz LA, Koretsky AP, Balaban RS (1987) FEBS Lett 221: 270
277. Lee RG, Lanir A, Clouse ME (1987) Invest Radiol 22: 479
278. Seo Y, Murakami M, Matsumoto T, Nishikawa H, Watari H (1987) Pflugers Arch 409: 343
279. Pieper GM, Murray WJ (1987) Biochem Med Metab Biol 38: 111
280. Murray PA, Blanck TJ, Rogers MC, Jacobus WE (1987) Anesthesiology 67: 649
281. Canioni P, Desmoulin F, Cozzone PJ (1987) Diabete Metab 13: 395
282. Murakami M, Seo Y, Watari H, Ueda H, Hashimoto T, Tagawa K (1987) Jpn J Physiol 37: 411
283. He MX, Wangler RD, Dillon PF, Romig GD, Sparks HV (1987) Am J Physiol 253: H1184
284. Lewandowski ED, Devous MDSr, Nunnally RL (1987) Am J Physiol 253: H1215
285. Kupriyanov VV, Lakomkin VL, Kapelko VI, Steinschneider AY, Ruuge EK, Saks VA (1987) J Mol Cell Cardiol 19: 729
286. Kingsley-Hickman PB, Sako EY, Mohanakrishnan P, Robitaille PM, From AH, Foker JE, Ugurbil K (1987) Biochemistry 26: 7501
287. Pissarek M, Grunder W, Keller T (1987) Biomed Biochim Acta 46: S564
288. Ambrosio G, Jacobus WE, Bergman CA, Weisman HF, Becker LC (1987) J Mol Cell Cardiol 19: 953
289. Schwalb H, Kushnir T, Navon G, Yaroslavsky E, Borman JB, Uretzky G (1987) J Mol Cell Cardiol 19: 991
290. Ugurbil K, Kingsley-Hickman PB, Sako EY, Zimmer S, Mohanakrishnan P, Robitaille PM, Thoma WJ, Johnson A, Foker JE, From AH (1987) Ann NY Acad Sci 508: 265
291. Clarke K, Willis RJ (1987) J Mol Cell Cardiol 19: 1
292. Fuchinoue S, Teraoka S, Tojimbara T, Nakajima I, Honda H, Ota K (1988) Transplant Proc 20: 953
293. Murphy E, Gabel SA, Funk A, London RE (1988) Biochemistry 27: 526
294. Wexler LF, Lorell BH, Momomura S, Weinberg EO, Ingwall JS, Apstein CS (1988) Circ Res 62: 766
295. Auffermann W, Wu S, Parmley WW, Higgins CB, Sievers R, Wikman-Coffelt J FASEB J 256

296. Shulman GI, Rothman DL, Chung Y, Rossetti L, Petit WA Jr, Barrett EJ, Shulman RG (1988) J Biol Chem 263: 5027
297. Sako EY, Kingsley-Hickman PB, From AH, Ugurbil K, Foker JE (1988) J Surg Res 44: 430
298. Ochi S, Ogawa Y, Imai H, Nakazawa M, Imai S (1988) J Cardiovasc Pharmacol 11: 326
299. Pisarenko O, Studneva I, Khlopkov V, Solomatina E, Ruuge E (1988) Biochim Biophys Acta 934: 55
300. Hall JE, Mackenzie NE, Mansfield JM, McCloskey DE, Scott AI (1988) Comp Biochem Physiol 89: 679
301. Hoekenga DE, Brainard JR, Hutson JY (1988) Circ Res 62: 1065
302. Neubauer S, Hamman BL, Perry SB, Bittl JA, Ingwall JS (1988) Circ Res 63: 1
303. Sako EY, Kingsley-Hickman PB, From AH, Foker JE, Ugurbil K (1988) J Biol Chem 263: 10600
304. Katz LA, Koretsky AP, Balaban RS (1988) Am J Physiol 255: H185
305. Hoerter JA, Lauer C, Vassort G, Gueron M (1988) Am J Physiol 255: 192
306. Davis DG, Murphy E, London RE (1988) Biochemistry 27: 3547
307. Wolfe CL, Gilbert HF, Brindle KM, Radda GK (1988) Biochim Biophys Acta 971: 9
308. Morban E, Kitakaze M, Chacko VP, Pike MM (1987) J Circ Res 63: 673
309. Seo Y, Steward MC, Mackenzie IS, Case RM (1988) Biochim Biophys Acta 971: 289
310. Chatham J, Gilbert HF, Radda GK (1988) FEBS Lett 238: 445
311. Auffermann W, Camacho SA, Wu S, Litt L, Parmley WW, Higgins CB, Wikman-Coffelt J (1988) Magn Reson Med 8: 58
312. Authier B (1988) Magn Reson Med 8: 80
313. Thoma WJ, Ugurbil K (1987) Magn Reson Med 893: 225
314. Kapelko VI, Kupriyanov VV, Novikova NA, Lakomkin VL, Steinschneider AY, Severina MY, Veksler VI, Saks VA (1988) J Mol Cell Cardiol 20: 465
315. Lee RG, Clouse ME, Lanir A (1988) Liver 8: 337
316. Janes N, Clemens JA, Glickson JD, Cameron JL (1988) Adv Alcohol Subst Abuse 7: 213
317. Whitman GJ, Kieval RS, Brown J, Banerjee A, Grosso MA, Harken AH (1989) Surgery 105: 100
318. Takami H, Furuya E, Tagawa K, Seo Y, Murakami M, Watari H, Matsuda H, Hirose H, Kawashima Y (1988) J Biochem 104: 35
319. Allis JL, Snaith CD, Seymour AM, Radda GK (1989) FEBS Lett 242: 215
320. Borchgrevink PC, Bergan AS, Bakoy OE, Jynge P (1989) Am J Physiol 256: H195 h
321. Murphy E, Steenbergen C, Levy LA, Raju B, London RE (1989) J Biol Chem 264: 5622
322. Adam WR, Craik DJ, Kneen M, Wellard RM (1989) Clin Exp Pharmacol Physiol 16: 33
323. Fuller BJ, Busza Al, Proctor E (1989) Cryobiology 26: 112
324. Fisher MJ, Dillon PF (1989) Magn Reson Med 9: 53
325. Burstein D, Litt HI, Fossel ET (1989) Magn Reson Med 9: 66
326. Nedelec JF, Capron-Laudereau M, Adam R, Dimicoli J, Gugenheim J, Patry J, Pin ML, Fredj G, Bismuth H, Lhoste JM (1989) Transplant Proc 21: 1327
327. Pissarek M, Grunder W, Keller T, Goos H, Mest HJ, Krause EG (1989) Biomed Biochim Acta 47: S121
328. Watters TA, Bouchard A, Wu ST, Parmley WW, Wikman-Coffelt J (1988) Heart Vessels 4: 128
329. Tanaka A, Chance B, Quistorff B (1989) J Biol Chem 264: 10034
330. Kupriyanov VV, Lakomkin VL, Steinschneider AY, Severina MY, Kapelko VI, Ruuge EK, Saks VA (1989) Biomed Biochim Acta 20: 1151
331. Thoma WJ, Ugurbil K (1989) Am J Physiol 256: G949
332. Zimmer SD, Ugurbil K, Michurski SP, Mohanakrishnan P, Ulstad VK, Foker JE, From AH (1989) J Biol Chem 264: 12402
333. Sauer FD, Kramer JK, Forester GV, Butler KW (1989) Biochim Biophys Acta 1004: 205
334. Steward MC, Seo Y, Case RM (1989) Pflugers Arch 414: 200
335. Sakai H, Yoshiyama M, Teragaki M, Takeuchi K, Takeda T, Ikata M, Ishikawa M, Miura I (1989) Osaka City U Life Sci 45: 217
336. Weiss RG, Chacko VP, Glickson JD, Gerstenblith G (1989) Proc Natl Acad Sci USA 86: 6426
337. Jelicks LA, Gupta RK (1989) J Biol Chem 264: 15230
338. Endre ZH, Allis JL, Ratcliffe PJ, Radd GK (1989) Kidney Int 35: 1249
339. Jeffrey FM, Storey CJ, Nunnally RL, Malloy CR (1989) Biochemistry 28: 5323
340. Chatham JC, Seymour AL, Harmsen E, Radda GK (1989) Cardiovasc Res 22: 833
341. Wunderlich CC, Janes NJ, Samphilipo MA Jr, Chacko VP, Anderson JH (1989) Invest Radiol 34: 619
342. Weiss RG, Chacko VP, Gerstenblith G (1989) J Mol Cell Cardiol 21: 469

343. Endre ZH, Allis JL, Radda GK (1989) Magn Reson Med 11: 267
344. Albert K, Sudmeier JL, Anwer MS, Bachovchin WW (1989) Magn Reson Med 11: 309
345. Forester GV, Saunders JK, Mainwood GW, Butler KW, Scott JR, Paradis H, Roy OZ, Deslauriers R (1989) Adv Exp Med Biol 248: 551
346. Fukuda H, Yasuda H, Shimokawa S, Tamura M (1989) Adv Exp Med Biol 248: 567
347. Bradamante S, Monti E, Paracchini L, Perletti G (1989) Biochem Biophys Res Commun 163: 682
348. Murakami M, Suzuki E, Miyamoto S, Seo Y, Watari H (1989) Pflugers Arch 414: 385
349. Chatham JC, Gilbert HF, Radda GK (1989) Eur J Biochem 184: 657
350. Ahr HJ, Karl W, Scherling D, Wunsche C (1989) Scand J Gastroenterol Suppl 164: 52
351. Henze E, Schoser K, Lietzenmayer R, Schnur G, Sauer M, Kunz R, Clauser M, Adam WE (1989) Z Kardiol 78: 553
352. Fukuda H (1989) Hokkaido Igaku Zasshi 64: 466
353. Stefenelli T, Wu ST, Parmley WW, Mason DT, Wikman-Coffelt J (1989) Am Heart J 118: 1228
354. Blum H (1989) Circ Shock 29: 291
355. Ohsuzu F, Yanagida S, Sakata N, Nakamura H (1990) Jpn Circ J 53: 1138
356. Wikman-Coffelt J, Auffermann W (1989) Invest Radiol 24: 976
357. Kitakaze M, Marban E (1989) J Physiol 414: 455
358. Neubauer S, Ingwall JS (1989) J Mol Cell Cardiol 21: 163
359. Kirkels JH, van Echteld CJ, Ruigrok TJ (1989) J Mol Cell Cardiol 21: 1209
360. Brainard JR, Hutson JY, Hoekenga DE, Lenhoff R (1989) Biochemistry 28: 9766
361. Gupta RK, Dowd TL, Spitzer A, Barac-Nieto M (1989) Renal Physiol Biochem 12: 144
362. Dowd T, Barac-Nieto M, Gupta RK, Spitzer A (1989) Renal Physiol Biochem 12: 161
363. Allis JL, Endre ZH, Radda GK (1989) Renal Physiol Biochem 12: 171
364. Clarke K, Sunn N, Willis RJ (1989) J Mol Cell Cardiol 21: 1315
365. Kingsley-Hickman PB, Sako EY, Ugurbil K, From AH, Foker JE (1990) J Biol Chem
366. Watts JA, Norris TA, London RE, Steenbergen C, Murphy E (1990) J Cardiovasc Pharmacol 15: 44
367. Koretsune Y, Marban E (1990) Am J Physiol 258: H9
368. Cohen Y, Chang LH, Litt L, Kim F, Severinghaus JW, Weinstein PR, Davis RL, Germano I, James TL (1990) J Cereb Blood Flow Metab 10: 277
369. Kemp PR, Seymour AM, Bailey IA, Radda GK (1990) Biochim Biophys Acta 1051: 293
370. Seymour AM, Chatham JC, Radda GK (1990) J Mol Cell Cardiol 22: 131
371. Vander Elst L, Goudemant JF, Mouton J, Chatelain P, Van Haverbeke Y, Muller RN (1990) J Cardiovasc Pharmacol 15: 377
372. Nedelec JF, Capron-Laudereau M, Adam R, Patry J, Dimicoli JL, Bismuth H, Lhoste JM (1990) Transplant Proc 22: 492
373. Headrick JP, Willis RJ (1990) Can J Physiol Pharmacol 68: 110
374. Mekhfi H, Hoerter J, Lauer C, Wisnewsky C, Schwartz K, Ventura-Clapier R (1990) Am J Physiol 258: H151
375. Nedelec JF, Megnin F, Patry J, Jullien P, Lhoste JM, Dimicoli JL (1990) Comp Biochem Physiol 95: 505
376. Lewandowski ED, Johnston DL (1990) Am J Physiol 258: H1357
377. Seo Y, Murakami M, Suzuki E, Kuki S, Nagayama K, Watari H (1990) Biochemistry 29: 599
378. Cohen SM (1987) (ed) Physiological NMR spectroscopy: From isolated cells to Man Ann NY Acad Sci vol 508, New York
379. Gupta RK (1987) NMR spectroscopy of cells and organisms vols 1 and 2. CRC Press, Boca Raton
380. Cheng H-M Barnett PA, Gonzalez RG (1987) In: Burt CT (ed) Phosphorus NMR in biology Chap 9. CRC Press Boca Raton, p 209
381. Osbakken MD, Mitchell MD (1990) In: Marcus ML, Shelbert MR, Skorton JJ, Wolf GL (eds) Cardiac imaging: Principles and practice. WB Saunders, Philadelphia, in press
382. Fossel ET, Ingwall JS (1982) In: Cohen JS (ed) Noninvasive probes of tissue metabolism, Chap 7. Wiley-Interscience, New York, p 173
383. Fossel ET, Morgan HE, Ingwall JS (1980) Proc Natl Acad Sci USA 77: 3654
384. Toyo-oka T, Najayama K, Umeda M, Eguchi K, Hosada S (1985) Heart and Vessels (Supp 1) Spectroscopy p 50
385. Kantor HL, Briggs RW, Balaban RS (1984) Circ Res 55: 261
386. Katz LA, Swain JA, Protman MA, Balaban RS (1988) Am J Physiol 24: H189

387. Bernard M, Menasche P, Fontanarova E, Canioni P, Grousset C, Piwnica A, Cozzone P (1985) Clin Chim Acta 152: 43
388. Ingwall JS, Kobayashi K, Bittl JA (1987) In: Gupta RK (ed) NMR spectroscopy of cells and organisms. vol 2. CRC Press, Boca Raton, p 51
389. Cohen SM (1987) In: Cohen SM (ed) Physiological NMR spectroscopy: From isolated cells to Man Ann NY Acad Sci New York, vol 508, p 109
390. Kaplan O, Kushnir T, Navon G (1986) Proc 5th Ann Meeting of SMRM, Montreal. Quebec 3: 906
391. Evanochko WT, Ng CT, Glickson JD (1984) Mag Reson Med 1: 508
392. Evelhoch JL, Safareto SA, Nussbaum GH, Ackerman JJH (1986) Radiat Res 106: 122

# Individual Nuclei

The last topic of the volumes on 'In Vivo Magnetic Resonance Spectroscopy' is devoted to in vivo MRS of specific nuclei. It was not the purpose to provide a comprehensive review, but rather to present practical concepts for the application of in vivo MRS and to describe its potential and limitations.

Due to the development of suitable techniques for spatial localization, solvent suppression, and spectral editing, hydrogen-MRS has become a rather popular technique for in vivo MRS. A steadily increasing number of important endogeneous compounds such as lactate, creatine, choline, amino acids (glutamate, glutamine, *N*-acetylaspartate), inositol, and glucose can be assessed in intact tissue. In the first chapter of this section $^{1}H$ in vivo spectroscopy is reviewed by Williams.

Carbon-13 spectroscopy may be considered as a metabolic eye to tissue metabolism. With its natural abundance of 1.1% the nucleus lends itself to studies with isotopically enriched compounds and the fate of the labelled substrate can be monitored with high specificity. $^{13}C$ MRS thus provides relevant information on substrate turnover and reaction pathways in cells, perfused organs, whole animals, and also in humans. A variety of technical problems had to be solved. Proton decoupling, nuclear Overhauser enhancement, and polarization transfer techniques enhance the sensitivity of the method. Care has to be taken to avoid excessive radio frequency power deposition in tissue. Another major problem of $^{13}C$ MRS is localization due to the large chemical shift dispersion. These and other aspects related to clinical applications of $^{13}C$ MRS are discussed by Beckmann.

Fluorine-19 is not usually present in mobile compounds in tissues. $^{19}F$ MRS can thus be applied to study exogenous tracers or drugs in the body without the interference of endogenous signals. Well-known examples are fluoropyrimidine anticancer drugs (e.g. 5-fluorouracil), fluorinated blood substitutes, and fluorinated anesthetics. Methodological aspects and applications of $^{19}F$ MRS are discussed by Prior, Maxwell, and Griffiths.

Sodium-23 has generated considerable interest in MRI since the tissue Na concentration is roughly 150 mM, which—together with the 100% natural abundance of this nucleus—allows the measurement of Na images with reasonable spatial and temporal resolution. In MRS, $^{23}Na$ and $^{39}K$ are mainly utilized to study transport processes across membranes. The development of suitable shift reagents (e.g. DyTTHA) which are not able to cross the cell membrane made it possible to distinguish between intra- and extracellular Na and K signals. Other cations such as Ca may be assessed by use of chelating ligands which contain NMR reporter nuclei (e.g. $^{19}F$). The application of MRS to the study of ion transport processes in cells and organs is reviewed by Ingwall.

The largest number of in vivo MRS studies reported to date involve phosphorus-31 since the nucleus has a good NMR sensitivity and 100% natural abundance. Phosphorus-31 is present in a number of key metabolites associated with the cellular energy and phospholipid metabolism. Moreover, important physiological parameters like the tissue pH value or magnesium concentrations may be derived. The potential of $^{31}P$ MRS is reviewed by Rudin and Sauter.

# In Vivo Proton Spectroscopy. Experimental Aspects and Potential

**S. R. Williams**

Department of Biophysics, Hunterian Institute, The Royal College of Surgeons of England, 35–43 Lincoln's Inn Fields, London WC2A 3PN, UK

## Table of Contents



The biochemical and clinical information which $^{1}H$ NMR spectroscopy can give in vivo is described. The major metabolites which have been detected in vivo are considered in turn. The measurement of lactate in ischaemia and anoxia is discussed, and the complementary role of $^{31}P$ spectroscopy is emphasised. The use of *N*-acetylaspartate as a concentration standard, and as a possible marker of neuronal cells is considered. Changes in glutamate and glutamine can be observed following hyperammonaemia, and the use of animal models and the application of such measurements to human subjects is described. Other amino acids are also discussed: taurine, $\gamma$-aminobutyrate, aspartate, glycine, alanine, histidine, anserine and carnosine. Other compounds which can be detected by $^{1}H$ NMR in vivo are inositol, creatine and phosphocreatine, choline-containing compounds and fat. The future role of $^{1}H$ NMR spectroscopy in vivo is considered, and it is concluded that applications to the brain will remain a major area of interest, and that in the next few years the number of clinical investigations will increase. However, examples are also given of $^{1}H$ NMR of kidney, muscle and tumours. Finally, it is stressed that spectroscopy in vivo must be supported by parallel studies of isolated cells and tissues, and of biopsy samples and extracts in vitro.

NMR Basic Principles and Progress, Vol. 28

# 1 Introduction

Several chapters in this series (Vols. 26 and 27) have dealt with the technical problems of recording $^{1}H$ spectra in vivo. These problems are concerned with localisation strategies, the suppression of the water resonance, and picking out specific metabolite resonances from within a crowded spectrum. In this chapter

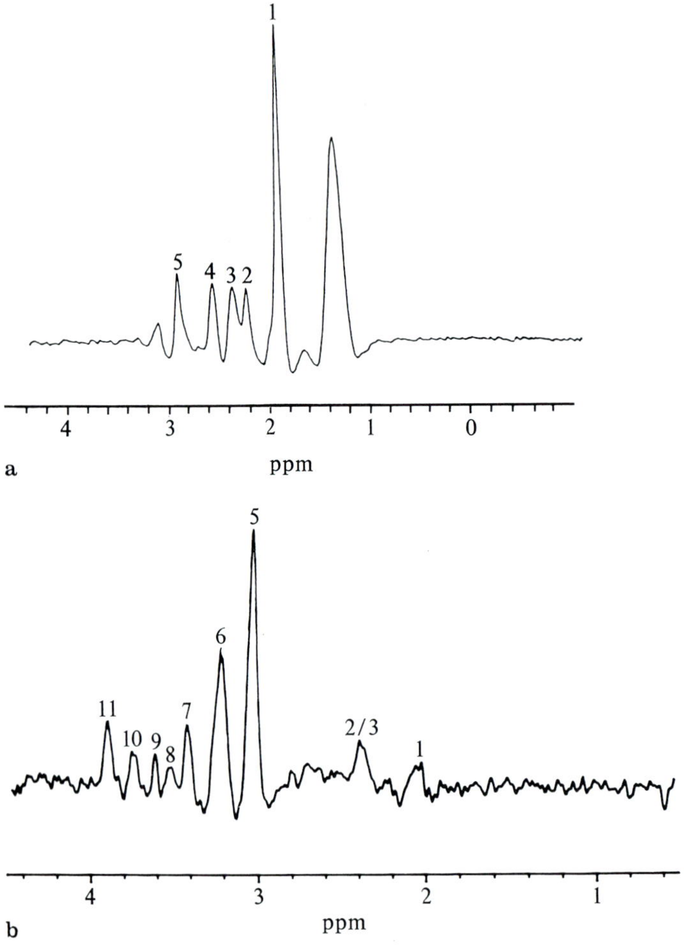


**Fig. 1.** $^{1}H$ NMR spectra recorded from rat brain in vivo. These spectra were recorded using a surface coil placed over the skull of an anaesthetised rat, at a frequency of 360 MHz in an 8.5 T vertical magnet. **a** was recorded using a $1\bar{3}3\bar{1}$–$2\bar{6}6\bar{2}$ solvent suppressed, spin echo sequence. The excitation maximum was set at 2.0 ppm. **b** was recorded using a $1\bar{3}3\bar{1}$–180 sequence, with the excitation maximum set at 3.03 ppm. In each case a total echo time of 120 ms was used (see Refs. 14, 48 for more details on the sequences). Assignment are as follows: *1*. *N*-acetyl aspartate, *2*. Glutamate, *3*. *N*-acetyl aspartate + glutamine, *4*. *N*-acetyl aspartate, *5*. Creatine + phosphocreatine, *6*. Choline containing compounds + taurine, *7*. Taurine, *8*. Inositol + glycine, *9*. Inositol, *10*. Glutamate + glutamine, *11*. Creatine + phosphocreatine. The unassigned peak at 1.2–1.4 ppm is from subcutaneous fat

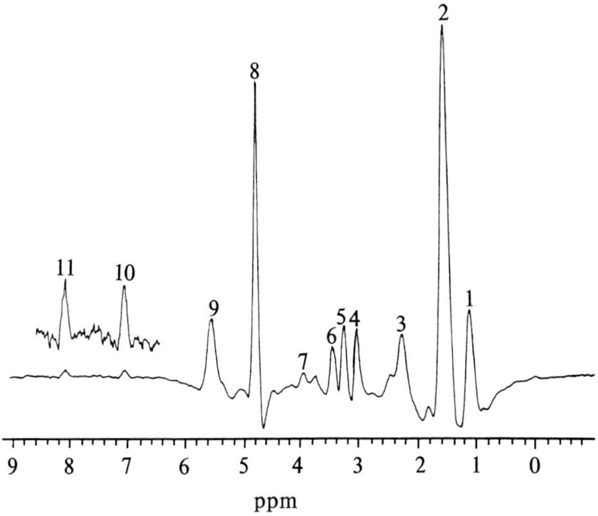


**Fig. 2.** $^1$H NMR spectrum recorded from rat leg in vivo. This spectrum was recorded using a saddle coil placed around the upper thigh of an anaesthetised rat at a frequency of 360 MHz in an 8.5 T vertical magnet. A Carr–Purcell–Meiboom–Gill spin echo sequence was used with a total echo time of 240 ms (see Ref. 15 for more details). Assignments are as follows: *1, 2, 3*. Fat, *4*. Creatine + phosphocreatine, *5*. Taurine + choline-containing compounds, *6*. Taurine, *7*. Creatine + phosphocreatine, *8*. Water, *9*. Fat, *10, 11*. Anserine

specific experimental protocols will not be discussed, but instead the chapter will concentrate on the metabolites which can be detected by $^1$H NMR, and what their significance is both in clinical terms, and in the advancement of basic biochemical and medical knowledge. There will be some discussion of technical aspects with regard to the problems of extrapolating from high field animal experiments to low field human investigations. Another aspect which will be covered in this chapter is the additional information which in vitro NMR can provide to aid the interpretation of $^1$H spectra recorded in vivo. The use of biopsy material, tissue extracts and NMR of isolated cell types will all be considered.

Much of the interest and excitement in $^1$H spectroscopy in vivo stems from the wide range of compounds that can in principle be studied, and the opportunity to look at a wide variety of metabolic pathways. However, the technical problems alluded to above have tended to restrict the applicability of the method, and it has been in the brain that $^1$H spectroscopy has been most useful. It has proved easier to record high quality $^1$H spectra from the brain than from other organs, partly because better field homogeneity can be achieved but also because the problems of fat suppression can be more easily solved by localisation and/or spectral editing in the brain than in other organs. Representative spectra of rat brain and skeletal muscle are shown in Figs. 1 and 2, respectively. The bulk of this chapter is therefore devoted to $^1$H NMR studies of cerebral metabolism, though mention is made of other organs where appropriate. The chapter is arranged by metabolite, so that all the information on a given metabolite is self-contained.

## 2 Lactate

The metabolite on which most attention has been focused for $^1$H spectroscopy is lactate. This is because lactate is a metabolically important marker of tissue ischaemia and hypoxia, since its concentration increases markedly in aerobic tissues which are deprived of $O_2$ for even short periods of time. In this respect $^1$H NMR offers similar information to $^{31}$P, since the increase in lactate is generally accompanied by a drop in high energy phosphates and is usually acknowledged as the cause of the fall in intracellular pH ($pH_i$). Given this association, is there any advantage to be gained from $^1$H NMR over $^{31}$P NMR? I believe $^1$H NMR on its own does offer some advantages over $^{31}$P NMR, both on technical and biological grounds, but ideally both nuclei should be used together, since the information they provide is complementary rather than identical.

The technical advantage of $^1$H NMR is its increased sensitivity, for the same number of nuclei it is about 15 times more sensitive that $^{31}$P NMR, and since the lactate methyl contains three protons, there is an effective increase in sensitivity per mole of 45. This increased sensitivity can be "invested" in improved signal-to-noise, reduced scanning time, or improved spatial resolution. This last factor is important in clinical spectroscopy since it can lead to reductions in the linear dimensions of voxel sizes of between three and four relative to a $^{31}$P spectrum, which allows spectroscopy to investigate anatomically meaningful regions of the human brain.

There is also a biological reason for measuring lactate either independently or in addition to recording $^{31}$P spectra: under certain situations changes in lactate, $pH_i$ and phosphorus metabolism do not occur simultaneously. This can be seen for example in the recovery from an acute ischaemic insult, in which ATP and $pH_i$ recover more quickly than lactate. In experiments using a model of cerebral ischaemia in the gerbil [1], ATP recovered back to control levels within 10 minutes of the end of a 30 minute ischaemic period, while $pH_i$ recovered within 15 minutes. Lactate, however, remained at its ischaemic level for about 5 minutes of the recovery period before declining, and returning to control within about 30 minutes. The importance of this observation in the general context of $^1$H NMR spectroscopy is that it indicates that changes in lactate and $pH_i$ need not accompany each other. In this specific example it supports the notion that normalisation of $pH_i$ occurs by processes other than lactate removal, and it has been suggested that $Na^+/H^+$ exchange may be involved [2]. A second example of the importance of lactate is provided by studies of cerebral hypoxia. Siesjo and colleagues [3] have shown using standard biochemical techniques that changes in lactate occur at slightly higher $O_2$ tensions than do disturbances in phosphorus metabolism. This effect can be demonstrated very elegantly by combined $^1H$ and $^{31}P$ NMR, in which a normal $^{31}P$ spectrum can be obtained at a $pO_2$ in the region of 25 mmHg, when a $^1H$ spectrum shows lactate elevated 4–5 fold above normal [4]. These examples are taken from acute, animal models of hypoxia and ischaemia but there is also evidence that lactate measurements will be important in human stroke, and may provide useful diagnostic or prognostic information some days after an infarct. Recently there have been reports of elevated lactate in infarcted regions of human brain at four days following a stroke [5], and in regions adjacent to an infarcted area at six months post-infarction [6]. The question arises as to whether the persistent lactate is simply the waste product of dead cells within the infarcted region, or whether the cells in this region are actively producing lactate by anaerobic glycolysis. This has been addressed by Shulman and colleagues [7] using $^1H$-observe, $^{13}C$-edited spectroscopy. They were able to demonstrate one month after an infarct, that $^{13}C$-label, administered as an infusion of labelled glucose, became incorporated into the lactate pool within the infarct, indicating active glycolytic metabolism in this cell population. These cells presumably no longer had properly functioning mitochondria and would be unable to make ATP, and it is interesting to note that $^{31}P$ spectra of infarcted regions tend to appear normal after about one month [8], although it appears that the total signal may be reduced [9]. This suggests that all phosphorus metabolites, including $P_i$, may have been lost from these cells.

Taken together, these observations show that in addition to an increased NMR sensitivity for $^1H$ relative to $^{31}P$, there are also situations in which lactate is a more sensitive indicator of defects in energy metabolism than phosphorus containing compounds.

Measurements of lactate can also provide useful biochemical information in situations other than ischaemia or hypoxia. In the brain glycolysis can be stimulated independently of increased oxygen consumption, and it has been suggested that anaerobic glycolysis may be an important energy source when rapid activation of cells is required. $^1H$ NMR is well placed to study this non-invasivity, for example Prichard and colleagues have been able to demonstrate local increases of lactate in the visual cortex following photic stimulation [10]. This information complements that available from positron emission tomography, which has shown increased glucose utilisation under similar conditions. The NMR data indicate that at least a proportion of this extra glucose is being metabolised anaerobically. Related studies in animal models have also shown increases in lactate, and reductions in high energy phosphates, following electrical stimulation [11], or the induction of seizures by drugs [12]. The particular importance of the human data is that it emphasises that stimulation of anaerobic glycolysis can occur under purely physiological circumstances and is not necessarily linked with a pathological insult.

In some of the work discussed above, it proved sufficient simply to demonstrate that the lactate concentration was abnormal. In other studies however, and for general applicability it is necessary to be more quantitative, and this question has been addressed by a number of workers. Careful comparisons of lactate spectra recorded in vivo from animal brain, with measurements of lactate in vitro from extracts of the same samples reveal an apparent discrepancy between the lactate signal intensity relative to markers such as *N*-acetylaspartate (NAA), creatine or water, and its concentration in vitro relative to these same compounds [13, 14]. However, if proper account is taken of all the factors which affect signal intensity in vivo, including $T_1$, $T_2$, phase modulation and the spatial dependence of the pulse sequences used, then effectively all of the lactate in the brain is NMR visible [14].

The preceding discussion in this section has been concerned solely with lactate measurements in the brain. This partly reflects a personal bias of interest, but is also broadly representative of the bulk of the literature on lactate measurements in vivo by $^{1}$H NMR. However, lactate has been detected in skeletal muscle in animals [15] and in humans [16]. In principle the studies of muscle disease, and particularly mitochondrial myopathies, which have been carried out using $^{31}$P NMR spectroscopy in humans, and which have contributed much to the understanding of muscle pathology [17], would be complemented by simultaneous measurements of lactate. However, because of the very large signals from fat, lactate detection in human muscle is more problematic than in brain (in the head NMR-visible fat is only present subcutaneously, and its contribution can be eliminated effectively by localisation and/or spectral editing). I am unaware of any applications of $^{1}$H NMR spectroscopy of lactate in the investigation of human muscle disease.

In the kidney, lactate accumulation during ischaemia has been followed by $^{1}$H NMR in the rat [18], and as discussed above the increased sensitivity relative to $^{31}$P NMR may make this an important technique clinically to investigate renal failure due to hypoperfusion. Another potentially important application lies in the field of tumour monitoring. Many tumours are presumed to rely on glycolysis as a major source of energy, thus they may accumulate lactate intracellularly. In animal models higher levels of lactate than are present in normal tissue have been measured by $^{1}$H NMR [19]. Human astrocytomas have also been shown to contain more lactate than non-invaded brain [20], but at this stage it is unknown whether this reflects anoxia within the tumour, or the "normal" (i.e. well-perfused) metabolic status of the glioma.

## 3 Amino Acids and Related Metabolites

### *Background*

Amino acids are important intermediates in a number of metabolic pathways, as well as being required for protein synthesis. In addition some have specific functions in the central nervous system as neurotransmitters or neurotransmitter precursors. The nitrogen metabolism which these compounds are involved with includes pathways essential to life such as the urea cycle and ammonia detoxification. Since all the major tissues have a role to play in nitrogen metabolism, it might be hoped that non-invasive $^1$H NMR spectroscopy could help in understanding these pathways in vivo. However, this expectation has not been met except in the brain, and the following discussion relates mainly to studies of cerebral amino acids and related compounds.

The sensitivity and spectroscopic resolution of $^1$H NMR in vivo is such that compounds can be detected in principle at concentrations of 1 mM or greater. Reference to standard textbooks on cerebral biochemistry [21] indicates that the following amino acid metabolites may be amenable to study by $^1$H NMR in normal brain:

| Compound | Concentration (mmol/kg wet weight) |
|---|---|
| Glutamate | 8–12 |
| *N*-acetylaspartate | 5–10 |
| Glutamine | 2–6 |
| Taurine | 1–5 |
| Aspartate | 1–3 |
| $\gamma$-aminobutyrate | 1–2 |
| Glycine | 0.5–1.5 |
| Alanine | < 1 |

Additionally other compounds may be detected if their concentrations become elevated in certain diseases, including genetic disorders.

## *3.1 N-Acetylaspartate*

*N*-acetylaspartate (NAA) has probably received more attention than any of the other nitrogen metabolites detectable in a brain spectrum. Although its concentration is less than that of glutamate, its resonance is much more prominent in vivo, because of the singlet methyl resonance from the *N*-acetyl moiety at 2.02 ppm. In contrast the methylene resonances of glutamate display a complex coupling pattern giving rise to a much lower amplitude signal. Since the discovery of NAA in 1956 [22] there has been a considerable literature on its synthesis, distribution and possible function, which has recently been reviewed by Birken and Oldendorf [23]. Despite the considerable efforts of biochemists and neurochemists its role still remains obscure. It is localised exclusively to the central nervous system, but its concentration varies in different parts of the brain

[24]. To date, the NMR community has used the signal as a marker, either of concentration or of specific cell types, but in both cases some controversy remains. It has been used as a concentration standard in animal spectroscopy [14], but its use has been criticised because its concentration is not uniform throughout the brain [24]. Despite the validity of this objection, it does appear empirically that NAA can be used successfully to quantify lactate in animal brains [14]. The advantage of using NAA as a standard is that its concentration does not change during acute metabolic insults such as ischaemia or hypoxia [24].

The second major use of the NAA signal is as a "neuronal marker", since there is good circumstantial evidence that the compound is only found in neurones. For example, only human brain tumours of neuronal rather than glial origin were found to contain NAA by Nadler and Cooper [25]; and Koller et al. [26] have shown that lesions produced by specific neuronally toxic agents lead to reductions in NAA. Recently $^{1}H$ NMR has provided further supporting evidence from studies of human tumours in vivo [20, 27], studies of tumour extracts [28] and studies of extracts from cultured cells [29]. The tumour data are in agreement with Nadler and Cooper, in that NAA is reduced in astrocytomas. Interpretation of the in vivo data is hampered by partial volume effects, but NMR spectra of biopsy specimens are helpful as a firm histological diagnosis of cell type is possible; the studies of Gill et al. [28, 29] show a considerable reduction or a complete absence of NAA for high grade astrocytomas. The in vivo data are also supported by studies of isolated cell types cultured from specific neural populations, since the NMR spectra of cultured astrocytes have been shown not to contain any NAA [29]. It is likely that the use of NAA as a neuronal marker may prove to be important clinically, since many diseases can lead to the destruction of neurones. For example, in stroke it has been reported that the NAA level is reduced in spectra recorded four days following an infarct [5].

The interpretation of the NAA signal is complicated in children and immature animals by the large change in the concentration of NAA during development. In rat brain NAA increases from less than 1 mmol/kg wet weight at birth to its adult level of about 5 mmol/kg within 21 days [22, 30], and it is interesting to note that the onset of the increase in NAA coincides with the onset of myelination. It has been suggested that NAA may have a role to play in supplying acetyl groups for myelin synthesis [31] which would be consistent with its developmental time course. It would not, however, explain why the level remains elevated long after active myelin synthesis has ceased. Given the close temporal association between NAA and brain maturation, it may be that delayed or impaired neurodevelopment will be reflected in changes in the NAA signal, which may prove to be an important diagnostic tool.

A further caution should be placed on the interpretation of the NAA signal, since it has been suggested by some workers that other compounds contribute significantly to the signal in vivo at around 2.0 ppm [32]. In fact, there is no doubt that the $\beta$-$CH_2$ protons of glutamate do overlap the NAA-methyl signal at high magnetic fields, which are often used for animal studies [33], but of more interest is the question as to whether non-extractable (i.e. macro-molecular) compounds give an NMR signal in this region of the spectrum. There are large amounts of *N*-acetyl containing compounds in brain such as glycoproteins and gangliosides [21], and it will be important to determine in the next few years whether any of these molecules can generate NMR signals in vivo. Currently there are conflicting reports as to whether or not other compounds may give rise to signals detectable in vivo at 2.0 ppm [32, 34].

To conclude, the NAA signal has been used largely as an empirical marker in NMR studies, but these studies have added more evidence to the hypothesis that the compound is located predominantly in neurones. It is to be hoped that the renewed interest in NAA sparked by $^{1}H$ spectroscopy may lead to successful efforts to discover its true function.

## *3.2 Glutamate and Glutamine*

Glutamate is one of the major excitatory neurotransmitters in the central nervous system, and it has been implicated as a possible cause of irreversible ischaemic damage to neurones (the so-called "excitotoxic" hypothesis [35]). It is present in all cell types in the brain, though its concentration is higher in neurones. Glutamate metabolism is compartmented in brain, in the sense that there are two pools which are labelled at different rates by glucose [36], and $^{13}C$ NMR has been used to investigate this [37]. Although a detailed discussion of these experiments is outside the scope of this chapter, it will be seen later that advantage can be taken of the rapid labelling of glutamate in $^{1}H$- observe, $^{13}C$-edited studies.

Glutamate can be easily detected at the high magnetic fields used for animal spectroscopy in vivo, since it is present at high concentration, and gives rise to one signal, from the $\gamma$-$CH_2$ protons at 2.35 ppm, which can be resolved from other metabolites [33]. Unequivocal identification of glutamine is more difficult, as the $\beta$-and $\gamma$-$CH_2$ protons co-resonate with the $\beta$-glutamate and NAA methyl protons at 2.0–2.15 ppm, and with one of the $\beta$-methylene protons of NAA at 2.45 ppm, respectively. At the low fields used for clinical studies detection of glutamate is more problematic. The coupling pattern becomes more complex, distributing the resonance intensity over more spectral lines, and in a spin echo or stimulated echo experiment complex phase-modulation occurs which can lead to cancellation of signals as echo times are prolonged. The solution to this problem is to work with short echo times, which historically have not been possible because eddy currents destroy the spectrum, but recent data with improved gradient coils and compensation methods indicate that glutamate detection is feasible in clinical magnets [34].

Behar et al. [38] were the first to report measurements of glutamate and glutamine in vivo by $^{1}H$ NMR. During hypoglycaemia they showed that the concentrations of glutamate and glutamine both declined. This probably reflects the utilisation of the carbon skeleton of these amino acids as an energy source by conversion to pyruvate via 2-oxoglutarate and other tricarboxylic acid cycle intermediates. The close relationship between glutamate and tricarboxylic acid cycle activity can be elegantly exploited in vivo by $^{1}$H-observe, $^{13}$C-edited spectroscopy. The label from $^{13}$C-glucose becomes rapidly incorporated into glutamate, and by following the appearance of the label via $^{1}$H NMR spectroscopy the rate of tricarboxylic acid cycle activity can be estimated [39]. This type of experiment may contribute significantly to our understanding of the compartmentalisation of glutamate metabolism in vivo.

The other major metabolic pathway involving glutamate and glutamine that has been investigated in vivo by $^{1}$H NMR is ammonia detoxification. Ammonia is toxic to the brain at relatively low concentrations, and normally the level of blood ammonia is kept below 50 μM by the operation of the urea cycle in the liver. However, if liver function is compromised and blood ammonia rises, then detoxification takes place directly in the brain. Ammonia is incorporated into glutamine from glutamate by the enzyme glutamine synthetase, which is located in astrocytes [40]. Despite this process, hyperammonaemia is associated with severe neurological disturbances resulting in convulsions, coma and death. Elevated blood ammonia is thought to be one of the major causes of hepatic encephalopathy. A number of groups have investigated hyperammonaemia using various models, including acute liver failure [33], ammonia infusions [41, 42], hepatic ischaemia and urease treatment [42]. The results are in good agreement, in that with each model there is a fall in glutamate and concomitant increase in glutamine. $^{31}$P spectroscopy, in each case, failed to show any significant deviations from the normal spectra, despite clear signs of severely impaired neurological function in some of the models. This re-emphasises the point made in the discussion about lactate measurements, that under some circumstances the $^{1}$H spectrum is more sensitive to pathology than the $^{31}$P spectrum. As discussed above, these types of experiments are feasible using clinical spectroscopy systems. A study of a patient with hepatic encephalopathy has been reported [43], in which differences in the glutamate and glutamine region of the spectrum relative to a control were observed, though quantitative interpretation was not possible.

### 3.3 *Taurine*

After NAA and glutamate, taurine is probably the easiest amino acid to identify in $^{1}$H NMR spectra of brain. In muscle it generates the most prominent of the amino acid resonances [15]. Unfortunately its role in these tissues, rather like NAA, is not properly understood, so although it can be measured by NMR, there is little that has been done with the measurement. Unlike NAA, it has not assumed any empirical significance, since it is in too crowded a spectral region to be a good concentration standard, and in the brain it seems to be present in all cell types. Various functions have been proposed for taurine, including acting as an osmoregulator, a neuromodulator, an anti-oxidant, as a regulator of $Ca^{2+}$ and as a sink for biological sulphur [44]. It is beyond the scope of this chapter to consider taurine's biochemistry in any detail, but perhaps, as with NAA, $^{1}$H NMR in vivo will lead to a greater understanding of this compound's function.

Taurine and NAA account respectively, for a large proportion of the non-protein organic zwitterions in muscle and anions in the brain. It is perhaps surprising, after some 100 years of biochemistry, that we are still uncertain as to what such major constituents of tissue are doing!

## *3.4 Inborn Errors of Amino Acid Metabolism*

There is a wide variety of metabolic disorders which are associated with specific genetic defects and lead to increases in the amounts of particular metabolites. Many of these diseases are life-threatening in childhood and can lead to impaired mental and physical development. $^1$H NMR spectroscopy in vitro has been used to investigate many of these disorders [45], and offers the potential as a rapid diagnostic technique for the detection of abnormal metabolites in urine. Although considerably less sensitive than combined gas chromatography—mass spectrometry, which is usually used for such analysis, NMR has the advantage of being a very rapid technique with no requirement for sample pre-preparation. Applications of spectroscopy in vivo to such problems are few, and seem less likely to be clinically important, since the primary clinical problem is diagnosis which can be performed much more easily and cheaply than by NMR in vivo. Nevertheless, the feasibility of using $^1$H NMR in vivo to investigate amino acidopathies was demonstrated in 1985, when elevated levels of histidine were detected and quantified in a mouse model of histidinaemia [46]. Histidinaemia is itself a harmless disorder, but serves as an NMR model for phenylketonuria, since it is the amount of an aromatic amino acid which is elevated in each case. Subsequently, increased cerebral concentrations of phenylalanine have been reported in a rabbit model of phenylketonuria [47]. In humans, $^1$H NMR spectroscopy in vivo has demonstrated elevated levels of NAA in patients with Canavan's disease [48]. This disease has been shown to be associated with a deficiency in aspartoacylase, the enzyme which catalyses the hydrolysis of NAA [49].

The role of spectroscopy in vivo would seem to be in understanding how cerebral concentrations of abnormal metabolites are controlled and how they are related to the amounts in blood and urine. In suitable animal models, it would be interesting to know whether acute episodes of ketosis are necessarily associated with increased amounts of abnormal metabolites in the brain. The conventional methods of blood and urine analysis cannot answer these questions.

## *3.5 Other Amino Acids in Brain*

Reference to the table at the beginning of this section indicates that alanine, aspartate, $\gamma$-aminobutyrate and glycine may be detectable by spectroscopy in vivo, but there have been relatively few reports describing measurements of these compounds. They are all detectable in brain extracts [24] but, in vivo the reduced resolution hampers certain identfication of their assignments. Elevated alanine has been observed in one of the hyperammonaemia models described above [33], and is also reported to be increased relative to other metabolites in some human meningiomas [28]. During hyperammonaemia, the elevated alanine may represent a generalised disturbance of amino acid metabolism, but in the case of the meningioma, it may be that increased alanine is a feature of meningeal cells. Further work involving extract studies and cell culture will help to elucidate this.

Aspartate, $\gamma$-aminobutyrate and glycine are all important neurotransmitters and I would hope that in the future $^1H$ spectroscopy in vivo may be able to add to our understanding of their role in normal tissue and disease, but at the time of writing this hope remains unfulfilled. Aspartate and $\gamma$-aminobutyrate are difficult to identify in vivo, because their resonances either overlap or are close to more intense signals from other compounds present at higher concentration such as NAA, glutamate and creatine. Glycine gives rise to just a single resonance at 3.55 ppm, which is difficult to distinguish in vivo from one of the inositol resonances.

### 3.6 *Anserine and Carnosine in Muscle*

The dipeptides anserine and carnosine ($\beta$-alanyl methyl histidine and $\beta$-alanyl histidine respectively), can be detected in $^1H$ spectra of muscle. Carnosine can be detected in humans [16] and frogs [50] and anserine in rats [14]. They are unique in being the only aromatic resonances to have received much attention in vivo in normal tissue. Their importance lies in the use of the histidinyl protons as reporters of $pH_i$, since the chemical shifts are sensitive to pH in the physiological range. A careful comparison of rat muscle $pH_i$ calculated from either anserine or the $P_i$ resonance in a $^{31}P$ spectrum has revealed no significant differences between the two methods [51].

## 4 Creatine and Phosphocreatine

The N-$CH_3$ and N-$CH_2$ protons of creatine and phosphocreatine give rise to prominent resonances in the $^1H$ spectra of muscle and brain. In vivo it is not possible to distinguish the two compounds because the chemical shift difference is too small to be resolved ($< 0.02$ ppm). As a result there is no information to be gleaned from these signals on phosphocreatine metabolism in hypoxia or ischaemia, since the signal from the sum of creatine plus phosphocreatine (total creatine) remains constant. However, this does mean that the creatine signal can be used as a concentration standard, in a similar manner to that described above for NAA. Creatine and phosphocreatine are present in both neuronal and glial cells, hence the NAA/(total creatine) ratio is often used as a indicator of neuronal loss. However, some caution should be exercised in the interpretation of this ratio, since studies of extracts from astrocytoma biopsies have revealed significant variations in total creatine (28), varying from that expected for normal brain (10–12 mmol/kg wet weight) to non-detectable. The complete absence of creatine signal probably reflects necrotic tissue, but the interpretation of reduced total creatine levels requires more work.

## 5 Choline Containing Compounds

There is a prominent resonance in the $^1H$ spectra of most tissues at around 3.2 ppm. It is usually assigned to "choline" or "choline-containing compounds". In fact there are a number of resonances containing the —$N(CH_3)_3^+$ moiety which can contribute to this peak. In brain and muscle, choline itself, phosphorylcholine and glycerophosphorylcholine are major constituents, but also taurine and the N-$CH_2$ resonance of phosphorylethanolamine will contribute some intensity. In liver and kidney betaine is the major compound in this region of the spectrum giving rise to a resonance at 3.26 ppm. It is not possible in vivo to separate out the different contributions to the peak, and even in extracts run at very high magnetic fields there are difficulties distinguishing between phosphorylcholine and glycerophosphorylcholine from the $^1H$ spectrum alone.

In brain spectroscopy the choline peak (where choline is used as a shorthand for all the compounds listed above) is assuming some significance as a marker in various disorders. Just as the NAA/(total creatine) ratio changes in certain tumours, it also appears that the choline/(total creatine) ratio changes. In astrocytomas, for example, the NAA/(total creatine) ratio appeared significantly different in all grades of tumour relative to normal tissue, whereas the choline/(total creatine) ratio was significantly different in grade IV tumours relative to other grades or normal brain [28]. The biochemical interpretation of changes in the choline peak is hampered by the uncertainty as to its constituents, as there may be a change in the proportions of different compounds contributing to the peak, and a change in the total amount of such compounds. Nevertheless, with respect to tumours, it is worth noting that in a $^1H$ NMR study of intact Friend leukaemia cells, both phosphorylcholine and glycerophosphorylcholine increased considerably when the cells were stimulated to divide [52]. Choline and its metabolites are involved in the pathways of phospholipid synthesis and degradation, so their concentration may vary according to the cell's needs for membrane synthesis.

Choline containing compounds and betaine have been suggested to play an important role in osmoregulation in the kidney [53], and $^1H$ and $^{14}N$ NMR spectroscopy have been important in establishing this [54, 55]. The $^1H$ NMR studies have generally not been performed in vivo, but advantage has been taken of the improved resolution and signal-to-noise available from extracts. Such invasive studies could not easily be performed on man, however, and spectroscopy in vivo would be appropriate in this case. Indeed, the feasibility of monitoring changes in renal osmolytes by $^1H$ spectroscopy in man has already been demonstrated [56], and further work should contribute to our biochemical and physiological knowledge in this area.

## 6 Inositol

Cerdan et al. reported the detection of inositol by $^1H$ NMR in brain extracts [57], and the compound can also be detected in kidney extracts [55]. In vivo, it is a difficult compound to detect unequivocally since its resonances are close to, or overlap those of other metabolites present at similar concentrations. Additionally, the use of spin echo or stimulated echo sequences leads to phase-modulation of all the inositol signals, further complicating detection. There is, however, a considerable interest in inositol metabolites biochemically, because of the importance of the inositol phosphates in intracellular signalling pathways [58]. It is interesting to note that, despite its lower sensitivity, $^{31}P$ NMR, rather than $^1H$ NMR, has been used in vivo to investigate cerebral inositol metabolism [59]. Inositol 1-phosphate levels were shown to increase following pharmacological stimulation of cats pretreated with lithium. This emphasises an important point: that the increased sensitivity of $^1H$ NMR does not always lead to an increased "detectability" in vivo, particularly for those compounds in which each resonance is just from a single proton. Considering all the compounds discussed in this chapter, it is apparent that the easiest to detect have singlet methyl resonances. Very complex spin systems, such as glutamate are only detectable if present at high concentration. For these reasons, $^1H$ NMR has not been used in vivo to investigate inositol metabolism, but there is a clear potential if spectroscopic quality can be improved, possibly with the use of editing techniques.

## 7 Fatty Acids

Much effort in $^1H$ spectroscopy in vivo is usually directed to suppressing resonances from fat, in order to detect signals from metabolites such as lactate. However, fat metabolism is an important field in its own right, and one to which NMR spectroscopy can perhaps contribute. An area of potential interest is in the release of free fatty acids which occurs during and following cerebral ischaemia, possibly as a result of membrane damage [60]. Indeed, Vink et al. [61] have shown by using editing techniques, that a signal increase in the lactate region of the spectrum following experimental ischaemia can not all be assigned to lactate. This work is generally interpreted as a warning concerning the use of non-edited lactate spectra, but it can equally well be used as a starting point for investigations of fat metabolism.

The investigation of fat deposits, for example fatty liver which occurs in cirrhosis, is a possible application of spectroscopy in vivo, but this type of study is probably better carried out with NMR imaging, since the concentration of fat is so high. In neuromuscular disease, spectroscopy has been used to characterise and quantify fat within muscle [62]. The amount of fat was found to be considerably elevated in a variety of neuromuscular disorders, a finding which in principle could have been demonstrated by imaging alone. However, spectroscopy was also able to identify specific resonances from unsaturated fats which could not be detected in normal muscle. In this case there is a clear advantage to be gained from using spectroscopy.

## 8 Conclusion

In this chapter I have attempted to give an overview of the metabolic, biochemical and physiological information that $^1$H NMR spectroscopy in vivo can provide. It should be apparent that there are many areas in which such investigations have already made an impact—in particular in cerebral ischaemia, stroke and investigations of intra-cerebral tumours. This chapter has drawn on examples both from animal work and studies of human subjects. It is likely that in the future there will be more emphasis on the clinical information which $^1$H spectroscopy can provide, but it seems that this will continue to be backed up by sound experimentation using animal models. It is important to stress that spectroscopy in vivo will always require parallel studies in vitro (tissue extracts, biopsy samples and cells) to aid in the interpretation of the data. In complex heterogeneous systems, consisting of different cell types it will become important to know which cells are responsible for which NMR signals. This type of work is already in progress, and for example, provides support for the contention that NAA is located predominantly in neuronal cells.

Improvements in techniques and instrumentation are in my opinion, unlikely to increase the range of compounds which can be studied in vivo by $^1$H NMR, beyond those considered above. However, I would hope that technical improvements will enable more information to be obtained in vivo from the amino acid metabolites and inositol, and that it will become possible to measure these compounds routinely in human subjects. Furthermore, improved spatial resolution will allow physiologists and biochemists to ask meaningful questions about regional metabolite differences within an organ, and how these may be affected in disease.

The brain has been the most fruitful hunting ground for the $^1$H spectroscopist, and seems likely to remain so, but applications of the technique to other parts of the body will surely increase. Measurements of lactate, osmolytes and fatty acids in tissues other than the brain have all been described in this chapter, and provide a basis for further research.

Finally, this chapter cannot be considered to include all the important applications of $^1$H spectroscopy in vivo. Although the field is less than ten years old, the literature is already too large to be encompassed within a single chapter. It has been my intention to reflect the important aspects of the literature with which I am familiar, from a biochemical standpoint, and to point out possible areas of future research. There seems little doubt that in another ten years it will become impossible even to attempt to review the experimental aspects and potential of $^1$H spectroscopy in a single chapter.

## 9 References


1. Allen K, Busza AL, Crockard HA, Frackowiak RSJ, Gadian DG, Proctor E, Ross Russell RW, Williams SR (1988) J Cereb Blood Flow Metab 8: 816
2. Siesjö BK (1985) Prog Brain Res 63: 121
3. Siesjö BK, Wieloch T (1985) Br J Anaesth 57: 47
4. Allen K, Busza AL, Proctor E, Gadian DG, Crockard HA (1990) Proc 9th Ann Meeting SMRM, New York, USA 1028 (Abstract)
5. Bruhn H, Frahm J, Gyngell ML, Merboldt KD, Hänicke W, Sauter R (1989) Magn Reson Med 9: 126
6. Berkelbach van der Sprenkel JW, Luyten PR, Van Rijen PC, Tulleken CAF, Hollander JA (1988) Stroke 19: 1556
7. Rothman DL, Howseman AM, Graham GD, Lantos G, Petroff OAC, Brass LM, Fayad PB, Shulman GI, Prichard JW, Shulman RG (1990) Proc 9th Ann Meeting SMRM, New York, USA 107 (Abstract)
8. Walsh KMA, Gross B, Licht J, Levine SR, Glasberg M, Smith MB, Helpern JA, Bueri J, Gorell J (1988) In: Appel S (ed) Current neurology, vol. 8. Year Book Publishers, Chicago p 295
9. Bottomley PA, Drayer BP, Smith LS (1986) Radiology 160: 763
10. Prichard JW, Rothman DL, Novotny EJ, Petroff OAC, Avison MJ, Howseman AM, Hanstock CD, Shulman RG (1989) Proc 8th Ann Meeting SMRM, Works-in-progress, Amsterdam, Netherlands. 1071 (Abstract)
11. Prichard JW, Petroff OAC, Ogino T, Shulman RG (1987) Ann NY Acad Sci 508: 54
12. Petroff OAC, Prichard JW, Ogino T, Shulman RG (1986) Ann Neurol 20: 185
13. Chang LH, Pereira BM, Weinstein PR, Keniry MA, Murphy-Boesch J, Litt L, James TL (1987) Magn Reson Med 4: 575
14. Williams SR, Proctor E, Allen K, Gadian DG, Crockard HA (1988) Magn Reson Med 7: 425
15. Williams SR, Gadian DG, Proctor E, Sprague DB, Talbot DF, Young IR, Brown FF (1985) J Magn Reson 63: 406
16. Pan JW, Hamm JR, Rothman DL, Schulman RG (1989) Proc Natl Acad Sci USA 85: 7836
17. Radda GK, Bore PJ, Rajagopalan B (1984) Br Med Bull 40: 155
18. Terrier F, Lazeyras F, Posse S, Aue WP, Zimmermann A, Frey BM, Frey FJ (1989) Magn Reson Med 12: 114
19. Sotak CH, Freeman DM, Hurd RE (1988) J Magn Reson 78: 355
20. Brühn H, Frahm J, Gyngell ML, Mertboldt KD, Hänicke W, Sauter R, Hamburger C (1989) Radiology 172: 541
21. McIlwain H, Bachelard HS (1985) Biochemistry and the central nervous system, 5th edn, Churchill-Livingstone, London

22. Tallan HH, Moore S, Stein WH (1956) J Biol Chem 224: 41
23. Birken DL, Oldendorf WH (1988) Neurosci Behav Rev 13: 23
24. Petroff OAC, Ogino T, Alger JR (1988) J Neurochem 51: 163
25. Nadler JV, Cooper JR (1972) J Neurochem 19: 313
26. Koller KJ, Zaczek R, Coyle JT (1984) J Neurochem 43: 1136
27. Langkowski JH, Wieland J, Bomsdorf H, Leibfritz D, Westphal M, Offermann W, Maas R (1989) Magn Reson Imaging 7: 547
28. Gill SS, Thomas DGT, van Bruggen N, Gadian DG, Peden CJ, Bell JD, Cox IJ, Menon DK, Iles RA, Bryant DJ, Coutts GA (1990) J Comput Assist Tomog 14: 497
29. Gill SS, Small RK, Thomas DGT, Patel P, Porteus R, van Bruggen N, Gadian DG, Kauppinen RA, Williams SR (1989) NMR Biomed 2: 196
30. Bates TE, Williams SR, Gadian DG, Bell JD, Small RK, Iles RA (1990) NMR Biomed 2: 225
31. Patel TB, Clarke JB (1980) Biochem J 188: 163
32. Hanstock CD, Rothman DL, Howseman AM, Lantos G, Novotny EJ, Petroff OAC, Prichard JW, Shulman RG (1989) Proc 8th Ann Meeting SMRM, Amsterdam, Netherlands 442 (Abstract)
33. Bates TE, Williams SR, Kauppinen RA, Gadian DG (1989) J Neurochem 53: 102
34. Frahm J, Michaelis T, Merboldt KD, Bruhn H, Gyngell ML, Hänicke W (1990) J Magn Reson 90: 464
35. Rothman SM, Olney JW (1990) Trends Neurosci 10: 299
36. Van den Berg CJ, Krzalic LJ, Mela P, Waelsch H (1969) J Biochem 113: 281
37. Badar-Goffer RS, Bachelard HS, Morris PG (1990) J Biochem 266: 133
38. Behar KL, Den Hollander JA, Petroff OAC, Hetherington HP, Prichard JW, Shulman RG (1985) J Neurochem 44: 1045
39. Fitzpatrick SM, Hetherington HP, Behar KL, Shulman RG (1990) J Cereb Blood Flow Metab 10: 170
40. Martinez-Hernandez A, Bell KP, Norenberg MD (1977) Science 195: 1356
41. Fitzpatrick SM, Hetherington HP, Behar KL, Shulman RG (1989) J Neurochem 52: 741
42. de Graaf AA, Deutz NEP, Bosman DK, Chamuleau RAFM, de Haan JG, Bovee WMMJ (1991) NMR Biomed 4: 31
43. Kreis R, Farrow N, Ross BD (1990) Lancet 336: 635
44. Zelikovic I, Chesney RW (1989) New Protective Roles for Selected Nutrients, Alan R Liss, New York p 253
45. Iles RA, Chalmers RA (1988) Clin Sci 74: 1
46. Gadian DG, Proctor E, Williams SR, Cady EB, Gardiner RM (1986) Magn Reson Med 3: 150
47. Avison MJ, Herschkowitz N, Novotny EJ, Petroff OAC, Rothman DL, Colombo JP, Bachmann C, Shulman RG, Prichard JW (1990) Ped Res 27: 566
48. Grodd W, Kraegeloh-Mann I, Petersen D, Trefz FK, Harzer K (1990) Lancet 336: 437
49. Matalon R, Michals K, Sebesta D, Deanching M, Gashkoff P, Casanova J (1988) Am J Med Gen 29: 463
50. Yoshizaki K, Seo Y, Nishikawa (1981) Biochem Biophys Acta 678: 283
51. Gadian DG, Proctor E, Williams SR (1987) Ann NY Acad Sci 508: 241
52. Agris PF, Campbell ID (1986) Science 216: 1325
53. Balaban RS, Burg MB (1987) Kidney Int 31: 562
54. Balaban RS, Knepper MA (1983) J Am Physiol 245: C439
55. Gullans SR, Blumenfeld JD, Balaschi JA, Kaleta M, Brenner RM, Heilig CW, Herbert SC (1988) J Am Physiol 255: F626
56. Avison MJ, Rothman DL, Nixon TN, Long WS, Shulman RG (1990) Proc 9th Ann Meeting SMRM, Works-in-progress, New York, USA, p 1160 (Abstract)
57. Cerdan S, Parrilla R, Santoro J, Rico M (1985) FEBS Lett 187: 167
58. Berridge MJ, Irvine RF (1989) Nature (London) 341: 197
59. Renshaw PF, Schnall MD, Leigh JS Jr (1987) Magn Reson Med 4: 221
60. Rehncrona S, Westerberg E, Akesson B, Siesjo BK (1982) J Neurochem 38: 84
61. Vink R, McIntosh TK, Faden AI (1988) Magn Reson Med 47: 95
62. Barany M, Venkatasubramanian PN, Mok E, Siegel IM, Abraham E, Wycliffe D, Mafee MF (1989) Magn Reson Med 10: 210
63. Frahm J, Michaelis T, Merboldt KD, Hänicke W, Gyngell ML, Bruhn H (1991) NMR Biomed 4: 201
64. Gyngell ML, Michaelis T, Hörstermann D, Bruhn H, Hänicke W, Merboldt KD, Frahm J (1991) Magn Reson Med 19: 489
65. Bruhn H, Michaelis T, Merboldt KD, Hänicke W, Gyngell ML, Frahm J (1991) Lancet 337: 745
66. Kreis R, Ross BD (1991) Proc 10th Ann Meeting SMRM, San Francisco, USA 401 (Abstract)

**Note added in proof**

*Footnote to page 63*

It has recently been reported that N-acetylaspartylglutamate can be detected as a component of the NAA peak in human spectra [63].

*Addendum:* Detection of cerebral glucose by $^{1}$H NMR.

Since the chapter was finished a number of reports have appeared describing measurements of glucose in brain by $^{1}$H NMR. Under normoglycaemic conditions, brain glucose is less than 2 mM [21], and given its complex spectral pattern, it is surprising that it can be detected in vivo. Indeed, even in brain extracts glucose resonances are not usually assigned because of overlap with more intense signals from inositol, taurine and other metabolites. Nevertheless, a study using localised spectroscopy in rats has been reported in which resonances are assigned to glucose and a concentration of 1.1 to 3 mM calculated (64). Verification on extracts that the detected signals are indeed from glucose will be important. Despite this reservation, however, a significant feature of this study was the demonstration that the glucose spectrum is simplified at lower field strengths (2.35 T), which would explain why glucose has not been detected in rats at high fields. The difficulty associated with spectral overlap of glucose signals with other resonances can be overcome by the use of $^{13}$C-edited $^{1}$H spectroscopy combined with infusion of $^{13}$C-labelled glucose; this technique has been used successfully to measure glucose in the rat brain [41]. Finally, in diabetic hyperglycaemia, when blood glucose increases 2–3 times above normal, it has proved possible to detect glucose by $^{1}$H NMR in humans [65, 66]. The observation of glucose in patients with considerably elevated plasma glucose, lends support to the assignments proposed in the rat study described above. Glucose can therefore be added to the list of $^{1}$H NMR detectable compounds, and is likely to prove a fruitful avenue for further research.

# In Vivo $^{13}C$ Spectroscopy in Humans

**N. Beckmann**

MR Center and Biocenter of the University of Basel, Klingelbergstr. 70, CH-4056 Basel, Switzerland

## Table of Contents



Due to the low natural abundance of $^{13}C$ nuclei, to their unfavourable gyromagnetic ratio, and to the necessity of removing the $^{13}C{-}^{1}H$ couplings using double resonance techniques the development of in vivo $^{13}C$ NMR spectroscopy has been slower than its $^{1}H$ and $^{31}P$ counterparts. Nevertheless, in the past few years $^{13}C$ NMR has also been extended to in vivo applications in humans. Proton broadband decoupled natural abundance $^{13}C$ spectra of the human calf, the abdominal region, and the head can now be obtained routinely with a measurement time of about 20 min. This time resolution has encouraged the performance of studies concerning the intravenous administration of $^{13}C$-labeled substances as well. A review of the methodology of in vivo $^{13}C$ NMR spectroscopy for applications in humans is presented in this chapter.

NMR Basic Principles and Progress, Vol. 28

# 1 Introduction

As soon as the first $^{13}C$ nuclear magnetic resonance experiments were performed in 1957 [38, 47], it was realized that the direct observation of carbon nuclei had some advantages over the equivalent proton studies. Indeed, the enhanced effective resolution possible in $^{13}C$ NMR was soon recognized. Carbon resonances of organic compounds are found over a chemical shift range of about 200 ppm, compared to 10 ppm for proton nuclei. In many cases it is also possible to obtain narrower resonances in $^{13}C$ than in proton spectra. This arises from the fact that $^{1}H$ spectra are extensively spin-coupled, leading to broadened resonances for most proton resonances. Additional advantages of $^{13}C$ NMR relate to the direct observation of carbon-containing functional groups with no attached protons (e.g. carbonyls, nitriles), and the direct observation of carbon reaction sites of interest. However, the extreme difficulties experienced in the first $^{13}C$ NMR experiments due to low sensitivity and poor spectral resolution limited the measurements to highly soluble, low molecular weight materials, severely restricting the early application of $^{13}C$ NMR spectroscopy. Nevertheless, by the mid 1960s some research groups, such as those of D.M. Grant, P.C. Lauterbur, J.D. Roberts, and J.B. Stothers, succeeded in studying many classes of organic compounds with $^{13}C$ NMR techniques. Finally, the advent of pulsed NMR techniques [29] and of decoupling [30] made $^{13}C$ spectroscopy a routine tool for chemical and in vitro biochemical analysis.

Modern $^{13}C$-NMR studies of in vivo metabolism began in 1972 when Eakin et al. employed $^{13}C$ NMR to monitor the metabolism of [1–$^{13}C$] glucose in an eucaryotic cell system [28]. With the advent of superconducting magnets with a large horizontal bore the methodology was extended to investigate non-invasively the biochemistry of animals and humans in vivo. The first results on model animals demonstrated that proton decoupled natural abundance $^{13}C$ NMR spectra of good resolution and signal-to-noise ratio could be obtained in measurement times of less than 30 min [1, 2, 23, 26, 56, 58, 59, 61, 68, 73]. In recent years spectra of comparable quality and time resolution have also been obtained in humans [6–8, 20, 35, 41, 69, 70]. Utilization of $^{13}C$ enriched substances as non-radioactive tracers allowed the study of the citric acid cycle, of the synthesis and degradation of glycogen, and of the gluconeogenesis and ketogenesis in animals and humans by $^{13}C$ NMR. Applications of $^{13}C$ NMR in vivo have been reviewed recently by Cerdan and Seelig [24], by Bottomley [19], and by Seelig and Burlina [67].

In this contribution the methodology of in vivo $^{13}C$ NMR spectroscopy for applications to humans is discussed. Because of the characteristics of the $^{13}C$ nuclei (natural abundance of only 1.1%, splitting of the $^{13}C$ resonances due to coupling with the $^{1}H$ nuclei, large chemical shift dispersion), $^{13}C$ measurements of humans pose some difficulties, since:

- The total measurement time should be limited to a maximum of one hour:
- The high frequency power deposition arising from decoupling should not exceed the safety specifications for humans;

—The signals from the superficial adipose tissue are much more intense than those arising from the metabolites of interest;
—The spectral localization methods, based on $B_0$ gradients in combination with frequency selective pulses, which are currently used for $^{1}H$ and $^{31}P$-NMR are not directly applicable to $^{13}C$ nuclei.

Different strategies for circumventing these problems will be discussed in this chapter. Most important is the correct choice of surface coils since they provide the first and probably decisive step towards spectral localization. Next, we will discuss acquisition and decoupling schemes, which are considered primarily from the point of view of safety. Decoupling plays an important role in in vivo applications not only because it simplifies the spectral analysis but also because metabolite resonances may remain hidden beneath the intense methylene signals of fatty acids in coupled spectra. Signal localization and enhancement based on polarization transfer, heteronuclear proton editing, and chemical shift imaging constitute the next subjects. Finally, applications concerning the administration to humans of $^{13}C$-enriched substances, which act as non-radioactive tracers, are presented.

## 2 Coils

The challenge in natural abundance $^{13}C$ NMR studies of humans lies in obtaining proton-decoupled spectra of a well-defined anatomical volume of interest within measurement times of less than 30 min. In addition, no heating of the conductive biological tissue should arise during decoupling. Therefore, the coil system chosen for in vivo applications should combine high sensitivity and efficient decoupling without excessive power deposition.

Surface coils operated in the transmitter and/or receiver mode appear to be the most suited experimental arrangement for this purpose [6, 20, 35, 41, 69]. The properties of surface coils are well known and are discussed in detail in other sections of this volume:

—The high filling factor of surface coils yields a good signal-to-noise ratio leading to measurement times of only several minutes;
—Due to the inhomogeneous $B_1$ field distribution of a surface coil, the origin of the NMR signals is restricted to a half sphere underneath the surface coil and contained within approximately a coil radius, and thus some spatial localization is attained;
—When the surface coil is used in the transmitter mode the contribution of superficial signals from the adipose tissue can be reduced by proper adjustment of the RF pulse power.

Due to these properties, circular surface coils have been generally used for the excitation and detection of $^{13}C$ signals.

The $^{1}H$ coil, on the other hand, should allow shimming and acquisition of images from the region of interest, and also guarantee an efficient decoupling. Different geometrical arrangements have been suggested for accomplishing this goal. For instance, in [35] the use of a butterfly $^{1}H$ coil in the transmitter and receiver modes has been described; broadband decoupling was possible with this coil without violating the Food and Drug Administration (FDA) guidelines. Bottomley et al. reported the use of a figure-eight-shaped $^{1}H$ surface coil with a crossing point on the axis of the parallel $^{13}C$ circular coil [20]. For volume-selective $^{13}C$ spectroscopy applications of the Vosing technique, Knüttel et al. [43] used a bird-cage coil for the excitation, and a surface coil doubly tuned at the $^{1}H$ and $^{13}C$ frequencies for receiving. Several groups adopted a circular surface coil, which was coplanar and concentric with a smaller $^{13}C$ circular coil, for proton imaging and decoupling [6, 41, 75]. This configuration permitted broadband decoupled natural abundance $^{13}C$ spectra of humans to be obtained in about 20–30 min [6].

The difficulties in employing surface coils for decoupling arise from the inherent RF field inhomogeneity which leads to an inefficient use of the RF power for decoupling. For example, the maximum receiving sensitivity of a circular surface coil is attained at a depth of about 0.5 to 0.6 times the coil radius when using an excitation pulse of 180° at the coil center. In contrast, the decoupling field is at a maximum close to the coil and decreases by a factor of about 0.6 in the $^{13}C$-sensitive volume, assuming a double-tuned single-coil probe [51]. Thus, approximately only 36% of the energy deposited in the sample is actually used for decoupling. In addition, cyclic sequences for decoupling tend to be inefficient in the heterogeneous RF field of a surface coil since an accurate setting of pulse flipangles is generally required. Nevertheless, it has been demonstrated experimentally that a configuration consisting of two coaxial and coplanar $^{13}C\{^{1}H\}$ coils, in which the proton coil is distinctly larger than the $^{13}C$ coil, allows the use of cyclic sequences due to a rather homogeneous $^{1}H$-decoupling field in the $^{13}C$-sensitive volume [6, 75]. The coplanar and concentric surface coil arrangement has also been proposed for the generation of homogeneous excitation $B_1$ fields [74], and for localization purposes [15]. By properly selecting the coil dimensions Mispelter et al. [51] obtained an optimized $^{13}C$-$\{^{1}H\}$ surface coil system, for which the maximum homogeneous decoupling field region was as close as possible to the volume of maximum $^{13}C$ sensitivity. Their geometric constraints were:

$$\frac{d_\omega}{2d_C} < 0.1 \qquad (1)$$

and

$$1.4 \leqq \frac{d_H}{d_C} \leqq 2.0 \qquad (2)$$

where $d_\omega$ was the diameter of the wire, and $d_C$ and $d_H$ were the diameters of the $^{13}C$ and $^{1}H$ coils, respectively. For coil systems obeying these geometric constraints the decoupling coil did not perturb the carbon RF field distribution significantly, while at the same time the mutual inductance was rather constant and independent of the inner coil design [51]. A drawback of the two coplanar surface coil arrangement is the perturbation of the decoupling field of the outer proton coil by the RF of the inner $^{13}C$ coil. In some instances this could result in an inefficient decoupling probe. However, since the perturbing $^{13}C$-field decreases faster with depth than the proton decoupling field of the outer coil, in some sample space the magnitude of the decoupling field presents a flat maximum with depth, i.e. a null RF gradient. By choosing the dimensions of the coil system according to Eqs. (1) and (2) the region of null decoupling RF gradient is matched with the volume of maximum $^{13}C$ sensitivity. Thus, the use of a system of coplanar and concentric $^{13}C\{^1H\}$ surface coils seems to represent an adequate compromise between optimal decoupling efficiency and minimum RF power deposition. High flux regions close to the proton coil must however be avoided by the use of spacers between the coil system and the human body [6].

Spectra generated with two coplanar and concentric $^{13}C/^1H$ surface coils [6] are comparable in quality and sensitivity to those obtained with a combination of a circular $^{13}C$ surface coil and a butterfly coil [35]. The decoupling powers calculated for both systems of coils are within the safety guidelines imposed by the FDA (see Sect. 4-**Decoupling**).

## 3 Schemes for the Acquisition of $^{13}C$ Spectra

For the acquisition of $^{13}C$ spectra different approaches have been used. The simplest approach has been the pulse-acquire scheme, in which the excitation was performed by a rectangular RF pulse [6]. Suppression of fatty acid signals was achieved by a 200° pulse in the plane of the surface coil. This scheme was also used in combination with $^{13}C$ chemical shift imaging [7, 55]. Although simple, it was robust for ensuring reproducibility of the acquired spectra.

Jue et al. [41] collected $^{13}C$ spectra with a $1 - t - \bar{1}$ sequence tailored to suppress the intense lipid signals at 30.0 ppm and to excite optimally the $C_1$ glycogen region at 100.5 ppm. The repetition time of $\approx$ 26.0 ms for the acquisition of coupled spectra was optimized for the $C_1$ glycogen signal, whose $T_1$ and $T_2$ relaxation times are 240 ms and 30 ms, respectively [2]. For the acquisition of $^1H$ decoupled spectra the repetition time was increased to 80.0 ms. Since this approach was optimized for the $C_1$ signal of glycogen, the short acquisition time led to insufficient decoupling of the other resonances and to a rather poor spectral resolution.

Adiabatic pulses have also been applied to excite the $^{13}C$ nuclei. For instance, Heerschap et al. [35] used an adiabatic rapid half passage excitation pulse [14] of 2.7 ms duration for the acquisition of $^{13}C$ NMR spectra. To suppress signals from subcutaneous adipose tissue this pulse was preceded by a block pulse with a flipangle of about 50° in the centre of the coil followed by a dephasing gradient.

Werner et al. [78] applied numerically optimized adiabatic sin/cos excitation pulses [14, 76] to increase the signal-to-noise ratio at the $C_1$ signal of glycogen. A pseudo-selective adiabatic excitation pulse was used, generated by introducing a gap of several μs within a 3.0 ms sin/cos pulse. Under these conditions the spectra seem to be less sensitive to baseline and phase errors. In this excitation scheme the on- and off-resonance magnetization follows adiabatically the effective $B_1$-vector generated by the pulse and the resonance offset. During the gap, the on-resonance magnetization keeps its alignment with $B_1$, while the off-resonance magnetization has sufficient time to orientate perpendicular to $B_1$ due to chemical shift evolution. After the gap, the on-resonance magnetization completes its adiabatic flip to the transversal plane whereas the off-resonance magnetization remains spin-locked perpendicular to $B_1$, where it is randomized due to $B_1$ inhomogeneities. In the $^{13}C$ spectra the most prominent fatty acid signals at 30.0 ppm and 130.0 ppm could be suppressed by opposite sweeping pulses with gaps of 135 μs and 350 μs, respectively. The exact position of the gap within the pulse was determined by solving the Bloch equations numerically, whereas the length of the gap was optimized experimentally [78].

## 4 Decoupling

In order to utilize the full potential of in vivo $^{13}C$ NMR it is necessary to remove the $^1H$–$^{13}C$ multiplet splittings by proton decoupling during the acquisition. As a consequence, the sensitivity and spectral resolution are improved due to the collapse of both short and long range $^1H$–$^{13}C$ couplings. Although decoupling is performed routinely in in vitro measurements, applications to humans require additional precautions since the specific absorption rate (SAR) of the decoupling RF field could lead to tissue overheating.

To minimize the RF power deposition it is possible to decouple selectively particular protons using soft pulses. For instance, selective decoupling with low power pulses has been applied for improving the detection of the $C_1$ resonance of glycogen in human muscle and liver [41]. However, this approach is only useful if the interest is restricted to one particular resonance. Other interesting resonances which might be present in the spectra remain proton-coupled.

Under conditions of broadband proton decoupling all $^1H$–$^{13}C$ multiplets will collapse, perhaps helping to reveal additional resonances or to uncover resonances otherwise obscured by the intense triacylglycerol signals. The price paid for this gain in information is the increase in the RF power deposition. However, studies performed at magnetic fields of 1.5 T indicate that the $^1H$ spectral bandwidth is small enough to allow broadband decoupling, resulting in SAR values within the guidelines recommended by the Food and Drug Administration (FDA) [6, 35]. The present FDA guidelines state two levels of concern for local SAR of extremities and trunk: (A) SAR $\leq$ 8 W/kg or (B) localized RF heating not greater than 40 °C at extremities and 39 °C in the trunk [77, 79].

Proton decoupling with surface coils constitutes an additional problem because of the inhomogeneous RF field: despite positioning of the decoupling coil sufficiently away from the body, e.g. with spacers, power dissipation in that part of the body closest to the coil will always exceed power dissipation in farther distant regions. For this reason decoupling sequences such as Waltz-4 and Waltz-8 [48] have proven useful in reducing the required decoupling power

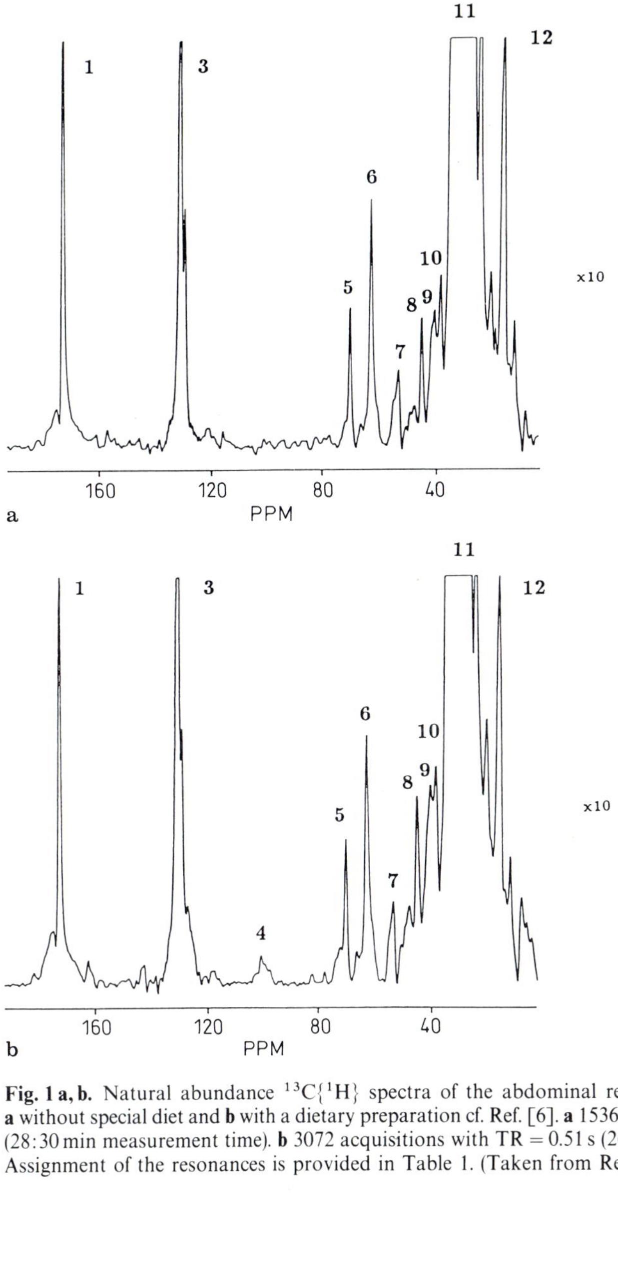


**Fig. 1 a, b.** Natural abundance $^{13}C\{^1H\}$ spectra of the abdominal region of a normal volunteer **a** without special diet and **b** with a dietary preparation cf. Ref. [6]. **a** 1536 acquisitions with TR = 1.11 s (28:30 min measurement time). **b** 3072 acquisitions with TR = 0.51 s (26:00 min measurement time). Assignment of the resonances is provided in Table 1. (Taken from Ref. [6])

level [6, 35]. Heerschap et al. determined an average local power absorption of ≦2.0 W/kg for superficial adipose tissue, of ≦3.2 W/kg for the muscle, and of ≦1.5 W/kg for the liver, considering a butterfly coil tuned at 63.6 MHz for proton decoupling with the Waltz-4 sequence [35]. Likewise the RF power deposition due to decoupling with Waltz-8 did not exceed 1.0 W/kg for applications on the brain, and 2.8 W/kg for the other organs [6, 7], when using a coplanar and concentric system of $^{13}C/^{1}H$ surface coils tuned at 16.0 MHz and 63.6 MHz, respectively [6, 7]. Decoupling was performed discontinuosly during 44.0 ms and 70.0 ms, respectively for measurements on the brain and other organs, i.e. intervals of 500 μs were intercalated between the rectangular $^{1}H$ decoupling pulses of length 500 μs for a 90° excitation. No temperature increase larger than 0.5 °C were observed—neither in measurements on conductive solutions nor at the body surface for measurement periods of about 30 min. Besides reducing the power deposition, the Waltz sequences provide efficient decoupling even though they are applied with surface coils. Using $^{13}C$ chemical shift imaging (CSI) Müller and Beckmann demonstrated that Waltz-8 leads to a homogeneous decoupling over a large volume [55]. This can be attributed to the fact that for a concentric and coplanar system of surface coils the maximum homogeneous region of the decoupling field overlaps to a large extent the volume of maximum $^{13}C$ sensitivity.

**Table 1.** Tentative assignments of $^{13}C$ resonances in the spectra of human calf, abdomen and head according to references [4]–[6], and [35]

| Nr. | Carbons | Chemical Shift (ppm) | Compounds |
|---|---|---|---|
| 1 | R–O–CO–R | 172.10 | fatty acid |
| 2 | $-N-C(NH_2)_2-$ | 157.1 | (P)creatine, arginine |
| 3 | –CH=CH– | ~128.4–130.1 | fatty acid |
| 4 | $C_1$ | 100.5 | glycogen |
| 5 | $C_2$ of glycerol backbone | 69.52 | triglyceride lipids |
| 6 | $C_1$ and $C_3$ of glycerol backbone | 62.36 | triglyceride lipids |
| 7 | $-CH_2-N-$; $N-(CH_3)_3$ | 54.0 | (P)creatine, GPC, (P)C, carnitine |
| 8 | $-CH_2-NH_2$ | 43.0–45.0 | PE (?) |
| 9 | β-carbon | 40.2 | AA |
| 10 | $-CH_3$ | 37.8 | creatine |
| 11 | $-CH_2-$ groups | 20–40 | fatty acid |
| 12 | $-CH_3$ | 14.4 | fatty acid |
| α | ? | ~85.0 | ? |
| β | ? | ~83.0 | ? |
| γ | ? | 77.5 | ? |
| δ | ? | ~72.0 | ? |

**Fig. 2 a–c.** Natural abundance $^{13}C\{^{1}H\}$ spectra of the human brain in vivo **a** and **b** normal volunteer, 1024 acquisitions with TR = 0.71 s (12:08 min measurement time). In **a**, the $^{13}C$ flipangle was 70°, and in **b**, 200° at the center of the surface coil. **c** Patient with type IIIA glycogen storage disease, 1280 acquisitions with TR = 1.1 s (23:50 min measurement time), flipangle 200°. (Taken from Ref. [6])

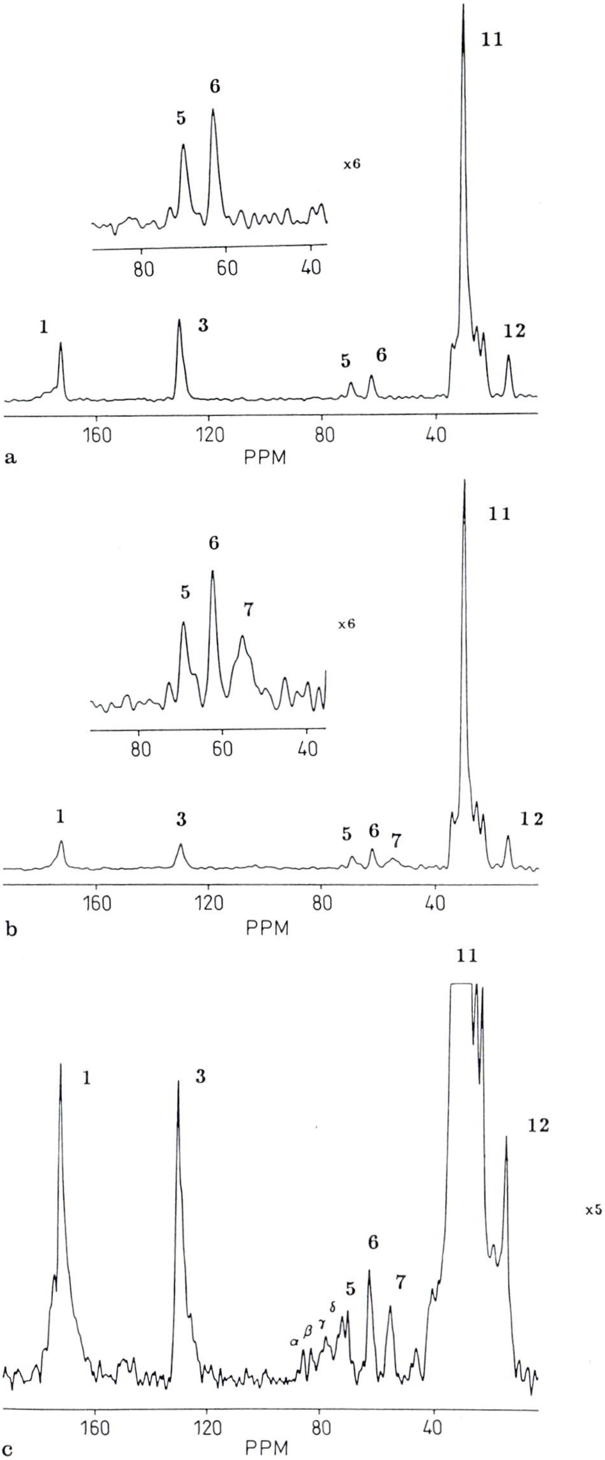
11
6
5
x6
80
60
40
1
3
5
6
12
160
120
80
40
a
PPM
11
6
5
7
x6
80
60
40
1
3
5
6
7
12
160
120
80
40
b
PPM
11
1
3
12
x5
6
α β γ δ 5
7
160
120
80
40
c
PPM

The $^{13}C$ spectra and $^{13}C$ chemical shift images presented in the following were all obtained with a system of concentric and coplanar $^{13}C/^{1}H$ surface coils in combination with a Waltz-8 decoupling scheme. Figure 1 shows spectra of the abdominal region of a normal volunteer without (Fig. 1a) and with (Fig. 1b) a special diet to increase the glycogen content [6]. The significant difference between the two spectra is the appearance of a broad glycogen $C_1$ resonance at 100.5 ppm in Fig. 1b. A tentative assignment of $^{13}C$ resonances from the spectra shown in this chapter is summarized in Table 1. This table indicates which metabolites have been detected in measurement times of less than 30 min [6, 35].

In vivo proton broadband decoupled $^{13}C$ spectra of the human head are summarized in Fig. 2 [6]. For a $^{13}C$ flipangle of about 70° (at the center of the surface coil) the spectrum exhibits essentially fat resonances (Fig. 2a). Increasing the $^{13}C$ flipangle to 200° leads to the appearance of a new resonance at 54.0 ppm which coresonates with (P)creatine/(P)choline (see Table 1). Natural abundance $^{13}C$ spectra were also obtained from a patient with type IIIA glycogen storage disease [6], showing additional resonances in the region from 71.0 to 85.0 ppm which could reflect sugar resonances (Fig. 2c).

Bottomley et al. reported proton-decoupled, Overhauser-enhanced natural abundance $^{13}C$ spectra obtained at 1.5 T with an arrangement consisting of a circular and a figure-eight-shaped surface coil [20]. Continuous decoupling and nuclear Overhauser enhancement (NOE) improved the $^{13}C$ sensitivity up to five-fold in the human leg, liver, and heart. It was estimated that continuous decoupling at 1.0 W RF power input resulted in local power deposition levels between 6 and 8 W/kg, which is just within the FDA guidelines of 8 W/kg [20, 77]. In spite of the gain in signal-to-noise due to the NOE, it is hence not recommended to apply continuous decoupling in vivo with surface coils since the power desposition could easily exceed FDA guidelines.

At higher magnetic fields, broadband decoupling will require even higher power levels because the necessary decoupling power increases with the square of the frequency range irradiated, and also because of the frequency dependence of RF absorption [62]. Indeed, Bomsdorf et al. demonstrated that broadband decoupling is severely restricted at 4.0 T since the RF-power deposition could result in SAR values above the maximum guideline values recommended by the FDA [16]. However,experiments on phantoms and volunteers using volume or surface coils for proton resonance showed that cw narrow-band decoupling might readily be performed at 4.0 T [16].

## 5 Signal Localization and Signal Enhancement Based on Polarization Transfer and Heteronuclear Proton Editing

As pointed out above, whole body systems in combination with surface coils operated in the transmitter and/or receiver mode have been an important experimental arrangement for in vivo $^{13}C$ NMR on humans. The surface coil approach is straight-forward to implement, has a high sensitivity, and allows the acquisition of spectra with a good resolution averaged over the whole measurement volume of the coil. In some cases the localization was improved with pulse sequences which take advantage of the $B_1$ field distribution of the surface coil (e.g. DEPTH pulses, which rely on extensive phase cycling of nonselective RF pulses [13]). As a result the sensitive volume of the RF coil was further restricted to regions characterized by specific $B_1$ field intensities. However, in many experimental situations an even more precise localization is required. For instance, surface coil $^{13}C$ spectra of skeletal muscle may be obscured by intense triglyceride signals from fat regions close to the surface coil. In this case, resonances from muscle tissue cannot be differentiated from similar signals of adipose tissue, and quantification becomes impossible.

For $^1H$ and $^{31}P$ in vivo spectroscopy powerful methods exist (e.g. ISIS [57], STEAM [32], PRESS [17]) which allow the image guided selection of a volume of interest for localized spectroscopy. Unfortunately, most of these techniques cannot be applied to $^{13}C$ due to the large chemical shift range of this nucleus. Frequency selective $^{13}C$ RF pulses in combination with $B_0$ gradients lead to a strong chemical shift dispersion in the spatial dimension so that each chemically shifted species has a different spatial origin [20]. This difficulty is less pronounced for $^1H$ and $^{31}P$ nuclei since their chemical shift ranges are much smaller. Other techniques which make use of RF excitation with a narrow excitation bandwidth (e.g. rotating frame spectroscopy [25]) lead to offset truncation in the spectra.

The application of polarization transfer and heteronuclear proton editing to in vivo $^{13}C$ NMR could, in principle, overcome these problems and lead to a better spectral localization and signal magnification. Let us first consider the polarization transfer from protons to carbons, which provides a maximum enhancement factor of

$$r = \frac{\gamma_H}{\gamma_C} \approx 4 \tag{3}$$

where $\gamma_H$ and $\gamma_C$ are the gyromagnetic constants of the $^1H$ and $^{13}C$ nuclei, respectively. This factor was calculated under the assumption that all motions fall in the short correlation time regime. An in vivo application taking advantage of this enhancement was reported in [66]. When applying a modified SINEPT sequence [40], whose performance is affected by the $T_2^*$ values of the protons bound to the $^{13}C$ nuclei, Saner et al. found an enhancement of only 2.1 for the $C_1$ resonance of glycogen at 100.5 ppm [66]. Due to relaxation effects, the enhancement factor of polarization transfer sequences in vivo could thus be smaller than predicted by Eq. (3).

An alternative way of enhancing the low intensity of $^{13}C$ signals is the measurement of protons bound to a $^{13}C$ nucleus of interest [63]. This approach aims at detecting $^{13}C$ with the sensitivity of protons, i.e., only signals from protons bound to $^{13}C$ nuclei are detected. Theoretically a maximum enhancement factor of

$$r = \left(\frac{\gamma_H}{\gamma_C}\right)^3 \approx 64 \tag{4}$$

is predicted, where $\gamma_H$ and $\gamma_C$ are the gyromagnetic ratios of the $^1H$ and $^{13}C$ nuclei, respectively [39]. A further enhancement proportional to the number of hydrogens bound to the carbon atom can be expected if the signal is acquired under decoupling conditions. The factor of Eq. 4 was estimated without taking the nuclear Overhauser effect and relaxation times into account.

## 5.1 *Localization Based on Polarization Transfer from $^1H$ to $^{13}C$*

Aue et al. [3] recognized that polarization transfer, in addition to being useful in signal enhancement and spectral editing, might also aid in generating spatially localized $^{13}C$ spectra. In this approach, schematically shown in Fig. 3a, the spatial selection was performed on protons and was followed by polarization

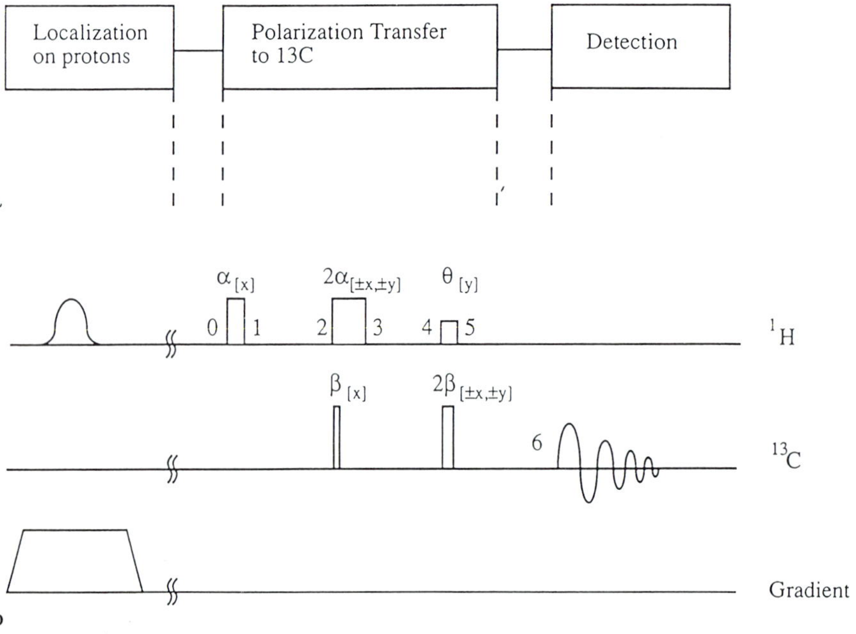


**Fig. 3a, b.** Pulse scheme for localized $^{13}C$ spectroscopy using $^1H$ volume selection followed by polarization transfer to $^{13}C$. The general scheme **a** may involve various localization and polarization transfer techniques. In this contribution scheme **b** is discussed where the localization on protons is performed with ISIS and the polarization transfer with DEPT. The ISIS inversion pulse is of the hypersecant type [64]. (Taken from Ref. [8])

transfer from the selected $^1H$ nuclei to $^{13}C$. The spatially localized and transferred magnetization was then observed in the $^{13}C$ space. Localization errors due to chemical shift dispersion were therefore minimized because the $^{13}C$ signals were localized via the volume selected by proton NMR.

Originally Aue et al. used homogeneous $B_1$ fields for protons and carbons [3]. In Ref. [8], a different approach is discussed where irradiation of both $^1H$ and $^{13}C$ nuclei is achieved with surface coils. Thus, the inherent advantages of surface coils are combined with the localization capability provided by the method discussed in [3]. The results presented in [8] were obtained with the pulse sequence shown in Fig. 3b, using coplanar and concentric $^{13}C$ and $^1H$ surface coils. The localization procedure was restricted to one dimension only, i.e. slices were selected parallel to the plane of the coils with a one-dimensional ISIS [57] scheme. In addition, the polarization transfer was performed with DEPT [60]. Comparison of the experimental results with theoretical simulations demonstrated that proton selection followed by polarization transfer to $^{13}C$ provided an excellent spatial localization of $^{13}C$ signals despite the use of surface coils in the transmitting and receiving mode [8].

Figure 4 shows proton decoupled natural abundance $^{13}C$ spectra from the human calf, where a layer of adipose tissue covers the gastrocnemius muscle. Figure 4a is the nonlocalized spectrum of the calf obtained with a $^{13}C$ pulse-acquire experiment, while in Fig. 4b the corresponding spectrum from a selected slice inside the muscle is displayed. A tentative assignment of the resonances based on references [4, 6, 35] was made in Table 1. In the localized spectrum (Fig. 4b) resonances {**a**}–{**e**} appear which were not resolved in the conventional surface coil spectrum (Fig. 4a). For instance, resonances {**b**} and {**c**}, which were tentatively assigned to $C_4$ and $C_{2,3,5}$ of glycogen, show up with more prominence in the DEPT-localized spectrum. The same is true for resonances {**d**}–{**f**}. The (P) creatine, (P)C resonance at 54.0 ppm has, however, its intensity diminished in the DEPT spectrum. Note that the carbonyl resonance at 172.0 ppm is absent in Fig. 4c since no protons are bound to carbonyls and thus no $^1H$ magnetization can be transferred to these $^{13}C$ nuclei.

The advantages of the present approach are:

— The selection is guided by a proton image. With selection techniques based solely on $B_1$ fields the selected region cannot be precisely delineated. The technique discussed here however enables a precise definition of the sensitive region via $^1H$ NMR, since the obtained $^{13}C$ magnetization is proportional to the $^1H$ magnetization preceding the polarization transfer (see [8]). This fact has two practical consequences: (a) selectivity may be verified with the more sensitive proton signals, for example by looking at $^1H$ projections or images of the selected region. $B_1$ techniques on the other hand demand the acquisition of $^{13}C$ signals in order to verify their spatial selectivity; (b) high flux signals, an important problem in $B_1$ localization techniques, are substantially reduced or even eliminated from the spectra.

— As the selection is performed on protons, different image-guided spatially selective sequences can be used preceding the polarization transfer. Selection

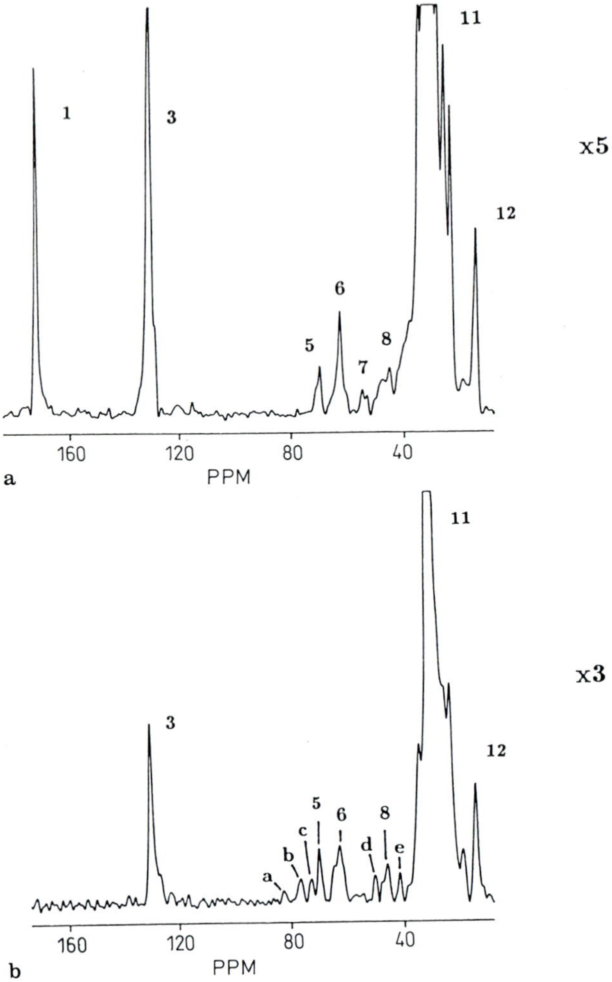


**Fig. 4 a, b.** Nonlocalized **a** and selected **b** spectra from the calf muscle of a normal volunteer. Measurement parameters: **a** 640 acquisitions with TR = 1.1 s (11:55 min); **b** 1536 acquisitions with TR = 1.1 s (29:00 min). (Taken from Ref. [8])

may be performed in one, two or three dimensions, and multifrequency excitation [52, 54] is equally possible. Polarization transfer may even be combined with techniques which selectively saturate the protons from a given sample region; the $^{13}C$ signal acquired after polarization transfer arises then from undisturbed sites in the sample.

— Polarization transfer provides an enhancement of signal intensity and allows editing of the $^{13}C$ signal.

— Chemical shift artefacts are minimized: errors in the localization due to the large chemical shift dispersion of the $^{13}C$ signals (approx. 200 ppm) are avoided in the present approach.

As a disadvantage of this technique it should be realized that signals from $^{13}C$ nuclei which relax very fast will hardly be detected because the polarization transfer sequence itself is rather long. Moreover, despite the possibility of shaping the volume according to the desired geometry, the acquired signal still represents an integrated response from the whole sample, since the volume selection is only introduced during the excitation period of the experiment and the acquisition of the signal is not local [53]. Therefore if contributions from outside the selected region persist, they will inevitably mix with those signals from the region of interest. However, inspection of Fig. 4 shows that: (i) the contribution of fat signals is reduced in the localized spectrum; (ii) some resonances in the spectral region from 40 ppm to 80 ppm are better characterized in the localized spectrum. These properties indicate that spatial selection on proton space combined with polarization transfer to $^{13}C$ may be a valuable tool for obtaining in vivo carbon spectra.

## 5.2 *Localization with the Vosing Technique*

Knüttel et al. introduced recently the Vosing technique for volume-selective $^{13}C$ spectroscopy [43]. This technique is based on cyclic polarization transfer (Cyclpot) and multiple-quantum filtering (MQF), and the signals to be detected are generally edited as coherence and spin-state transfer echoes. Field-gradients are applied for slice selection, for spoiling undesired coherences, and for balancing pairs of dephasing/rephasing intervals. In Cyclpot editing coherences resulting from a cyclic polarization transfer pathway are detected while all other resonances are suppressed. The proton magnetization is first transferred to $^{13}C$ coherences for all $^{1}H$ nuclei bound and properly coupled to $^{13}C$ nuclei. At this point, spoiler-gradients are applied. Coherences of coupled protons are selectively refocussed at a later stage of the sequence after having transferred their polarization back from the $^{13}C$ to the $^{1}H$ nuclei. The MQF line-editing on the other hand is based on the transfer of single-quantum $^{1}H$ coherences to heteronuclear multiple quantum coherences, which are then transferred back to single-quantum $^{1}H$ coherences. Echoes from all other coherence pathways, in particular from uncoupled protons, are suppressed. The Vosing technique has the advantage of being a single-scan editing procedure in contrast to the conventional two-scan subtraction sequences originally proposed for line editing [63]. It permits line-editing on a volume of interest to be carried out guided by an image.

An interesting in vivo application of the Cyclpot-Vosing technique was the detection of the $H_{1-4}$ resonance of liver glycogen [44], which occurs at 5.4 ppm in the proton spectrum. The Vosing technique has also been extended to imaging selectively protons coupled to $^{13}C$ nuclei [45].

Special attention must be given to relaxation times. Firstly, the acquired signals are affected by the $T_1$ relaxation times of the $^{13}C$ nuclei, which are longer than those of protons. Secondly, the acquired echoes are strongly affected by $T_2$ relaxation. Therefore, signals arising from metabolites with short $T_2$ relaxation times will be difficult to detect in vivo.

### 5.3 *Editing with the Poce Technique*

This technique is based on the multinuclear multipulse sequence described by Bendall et al. [12], which allows the observation of protons coupled to $^{13}C$ while eliminating signals of protons bound to $^{12}C$ nuclei. The sequence is basically a spin-echo sequence for the protons. In addition, an inversion pulse is applied to the carbons on alternate scans only. The inversion pulse has no effect on protons attached to molecules containing $^{12}C$, and signals from such protons are canceled by subtraction of alternate scans. This approach has been adapted by Hetherington et al. in order to generate proton-observed carbon-edited (Poce) spectra in vivo with a surface coil [31, 37]. Frequency-selective excitation was performed with rectangular composite pulses, minimizing the $B_1$ inhomogeneity of the surface coil. Phase cycling the RF pulses according to a DEPTH scheme [13] increased the spatial selectivity of the Poce technique.

Localized $^{13}C$-edited $^1H$ spectra of the human brain [65] were obtained using the Poce approach in combination with the ISIS technique [57]. In [65] edited ISIS spectra were obtained after the intravenous infusion of 99% enriched [1-$^{13}C$] glucose. By following the resonance of the proton bonded to the $^{13}C_4$ of glutamate at 2.3 ppm, the $^{13}C$ turnover time of the $C_4$ of glutamate was determined to be around 20 min. The results demonstrated that localized Poce spectroscopy can be used to measure regional glutamate turnover in the human brain.

## 6 $^{13}C$ CSI

In magnetic resonance studies the method of choice for generating metabolic maps is spectroscopic imaging (SI) [22]. In the past, SI has been applied in $^1H$ and $^{31}P$ studies demonstrating e.g. the heterogeneity of brain tumors [50] and of ischemic regions in humans [18]. Recently the technique was extended to $^{13}C$ and first experimental results on phantoms and in vivo on humans have demonstrated the potential of this technique [7, 55].

$^{13}C$ Spectroscopic Imaging has the following advantages [55]:

- The phase encoding with pulsed $B_0$ gradients does not introduce any chemical shift dispersion in the spatial dimension; spatial information is completely separated from the spectroscopic information.
- Since spatial resolution is only introduced via $B_0$ gradients, the $B_1$ field does not interfere with the localization mechanism. Therefore surface coils can be used.

— The low sensitivity of $^{13}C$ usually requires extensive signal averaging. Thus the phase encoding period does not prolong substantially the measurement time, and spatial mapping of metabolites is obtained without compromising the time resolution of the $^{13}C$ NMR experiment.

Spectroscopic Imaging is a multidimensional experiment providing spatial resolution in up to three dimensions. In [7, 55] the experiment was simplified

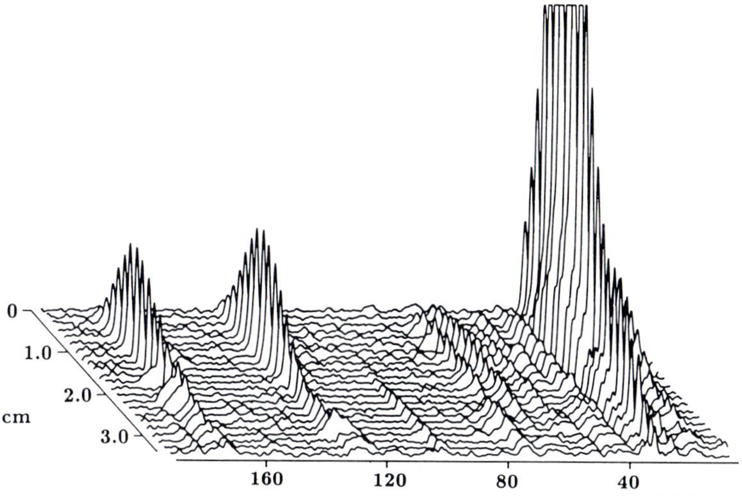


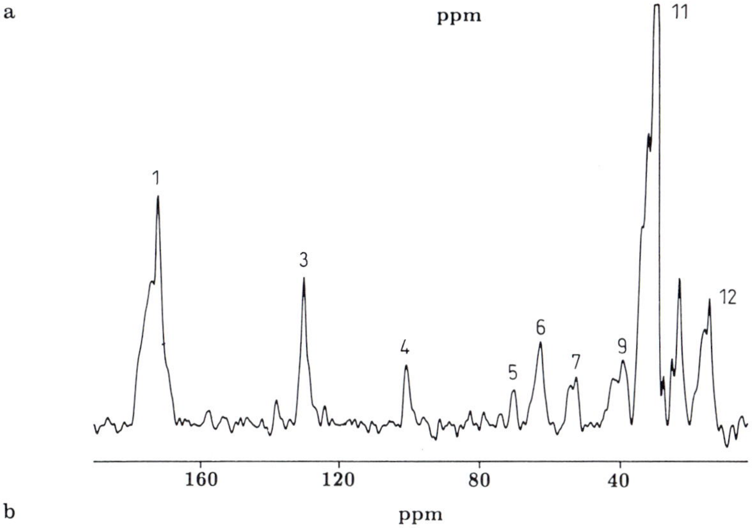


**Fig. 5 a.** Surface coil $^{13}C$ SI applied to the abdominal region of a patient with type IIIA glycogen storage disease. Decoupling with Waltz-8. Acq. time: 25.6 min, TR = 1.2 s. 64 phase encoding steps (−2.0 to 2.0 mT/m), 20 averages. Real part map, each spectrum consisting of 1024 points. **b** A spectrum taken out of the SI image shows clearly the glycogen signal at 100.5 ppm. Assignment of the resonances is provided in Table 1. (Taken from Ref. [7])

by reducing the dimensionality to one spatial and one spectroscopic axis. In this case the spectra were acquired along a single phase encoding gradient, i.e. each spectrum had its origin in a slice orthogonal to the direction of the phase encoding gradient. Therefore, when applied at the periphery of the human body, the slices closest to the coil selected mainly subcutaneous adipose tissue. Signals

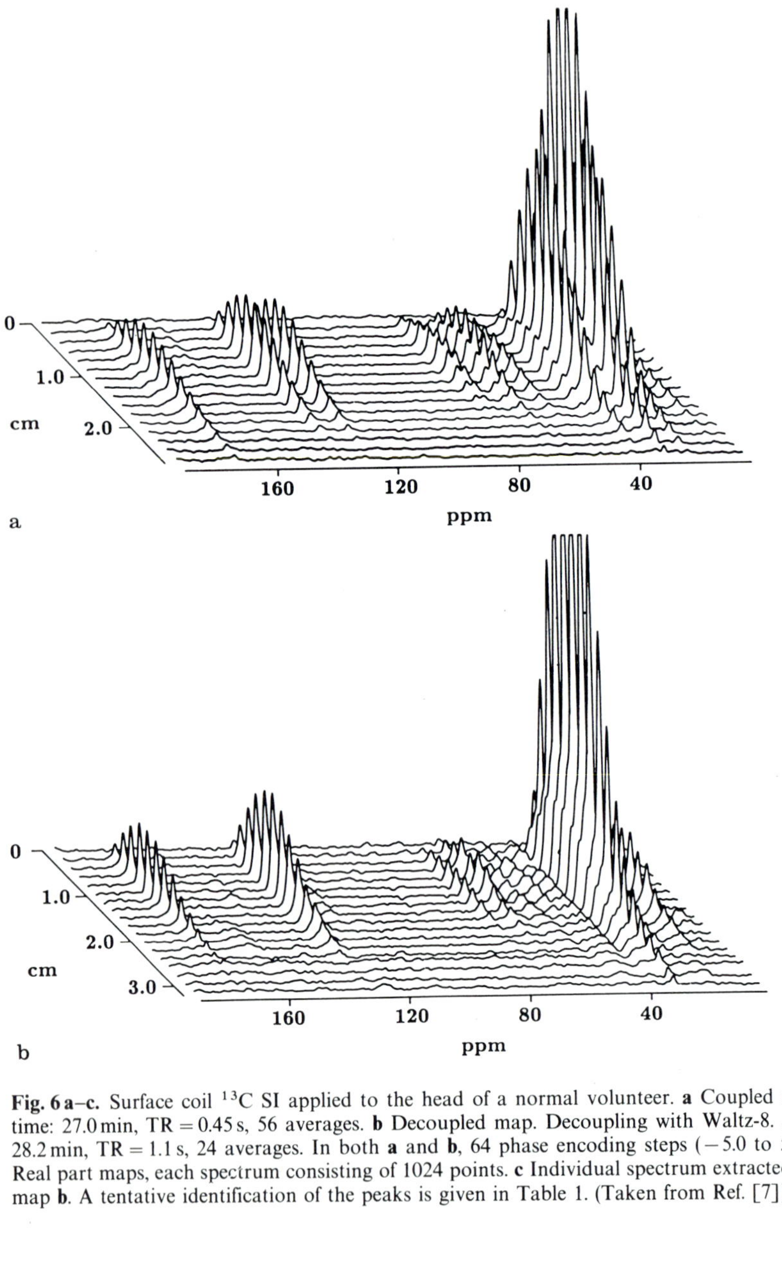


**Fig. 6 a–c.** Surface coil $^{13}C$ SI applied to the head of a normal volunteer. **a** Coupled map. Acq. time: 27.0 min, TR = 0.45 s, 56 averages. **b** Decoupled map. Decoupling with Waltz-8. Acq. time: 28.2 min, TR = 1.1 s, 24 averages. In both **a** and **b**, 64 phase encoding steps (−5.0 to 5.0 mT/m). Real part maps, each spectrum consisting of 1024 points. **c** Individual spectrum extracted from the map **b**. A tentative identification of the peaks is given in Table 1. (Taken from Ref. [7])

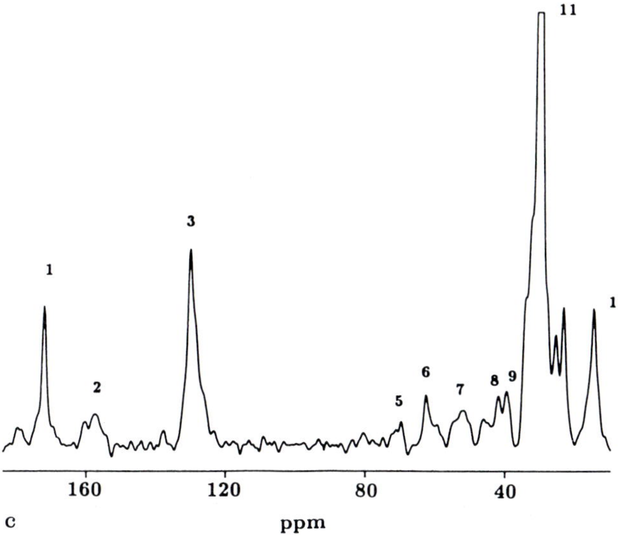


**Fig. 6.** (*Continued*)

from deeper lying organs could be found in the remaining sections of the spectroscopic image [7].

A $^{13}C$ SI map obtained from the abdominal region of a patient with type IIIA glycogen storage disease is displayed in Fig. 5a [6, 7]. The main feature in Fig. 5a is the $C_1$ resonance of glycogen at 100.5 ppm. In addition, the map shows resonances around 54.0 ppm and 40.0 ppm, which were tentatively assigned to (P) choline and $\beta$-carbons of aminoacids, respectively, according to Ref. [6]. An individual spectrum taken from the SI image clearly shows the glycogen signal at 100.5 ppm (Fig. 5b). Notice that the spectral resolution of this spectrum is comparable to that of the spectra resulting from a simple pulse-acquire experiment shown previously (see Figs. 1 and 2). Thus, taking into account the effect of relaxation during the time delay between excitation and acquisition in the SI experiment, quantification should also be possible.

Figure 6 shows $^{13}C$ SI results from measurements on the head of a normal volunteer. In this example the effect of proton decoupling is demonstrated. The map of Fig. 6a was obtained without decoupling with a simple pulse-phase encode-acquire $^{13}C$ experiment. In contrast, the acquisition of the data displayed in Fig. 6b was accompanied by broadband decoupling with a Waltz-8 sequence. Comparison of Figs. 6a (coupled spectra) and 6b (decoupled spectra) immediately shows the collapse of the *J*-multiplets in Fig. 6b. Note that the decoupling is very homogeneous despite the use of surface coils. The maps are dominated by fatty acid signals. Closer inspection of Fig. 6b reveals, however, additional resonances around 157.0 ppm ($C_3$ creatine), 54.0 ppm ($C_4$ creatine,

$C_5$ *sn*-glycerol 3-phosphorylcholine GPC), 47.0 ppm (unknown), 42.0 ppm ($C_2$ phosphorilethanolamine PE), and 38.0 ppm ($C_2$ creatine). These tentative assignments were based on Ref. [5]. Figure 6c displays an individual spectrum of the map of Fig. 6b.

## 7 Infusion Studies

The information content of a $^{13}C$ MR spectrum can be substantially increased if $^{13}C$-labeled substrates are administered as non-radioactive tracers. During the metabolism of the externally supplied $^{13}C$ labeled compounds (typically 90% enriched in $^{13}C$), the $^{13}C$ label is transferred to various intermediates which, in part, may become detectable with $^{13}C$ MRS. Hence metabolic pathways and fluxes may become amenable to an in vivo analysis. Until now, most in vivo studies with $^{13}C$-enriched substrates and $^{13}C$ MRS were restricted to animals, with particular emphasis on head and liver (for reviews see [24, 49]).

Studies on humans have also been reported, following two different approaches:

— The infusate was administered for a limited time period, followed by an analysis with in vitro NMR spectroscopy of blood plasma samples drawn at appropriate intervals. As examples the following studies can be cited: (i) determination of the glucose production in a human subject by continuous infusion of tracer levels of [U-$^{13}C$] glucose over a 4-h period and subsequent analysis of the plasma levels of tracer by in vitro NMR [21]; (ii) Gopher et al. estimated the hepatic fructose conversion to glucose by examination of $^{13}C$ NMR spectra of plasma glucose of normal and hereditary fructose-intolerant (HFI) children [34], after [U-$^{13}C$] fructose had been infused nasogastrically. Significantly lower values ($\approx$ 3-fold) for fructose conversion to glucose were obtained in HFI patients as compared to the controls. Additionally, the results suggested that there is a direct pathway from fructose to fructose 1,6-biphosphate, bypassing fructose-1-phosphate aldolase [34]; (iii) Kalderon et al. used [U-$^{13}C$] glucose for estimating the glucose carbon recycling in children with type I and type III glycogen storage disease (GSD) [42]. The results indicated that there is no glucose production via gluconeogenesis in type I GSD-patients, while in type III GSD-patients, gluconeogenesis was suggested as the major route for endogenous glucose synthesis [42].

— The NMR measurements were performed in vivo in combination with an infusion of $^{13}C$ labeled substances. For this category of experiments the following examples can be found in the literature: (a) Shulman et al. assessed the rate of incorporation of intravenously infused [1-$^{13}C$] glucose in the gastrocnemius muscle of normal subjects and of patients with non-insulin

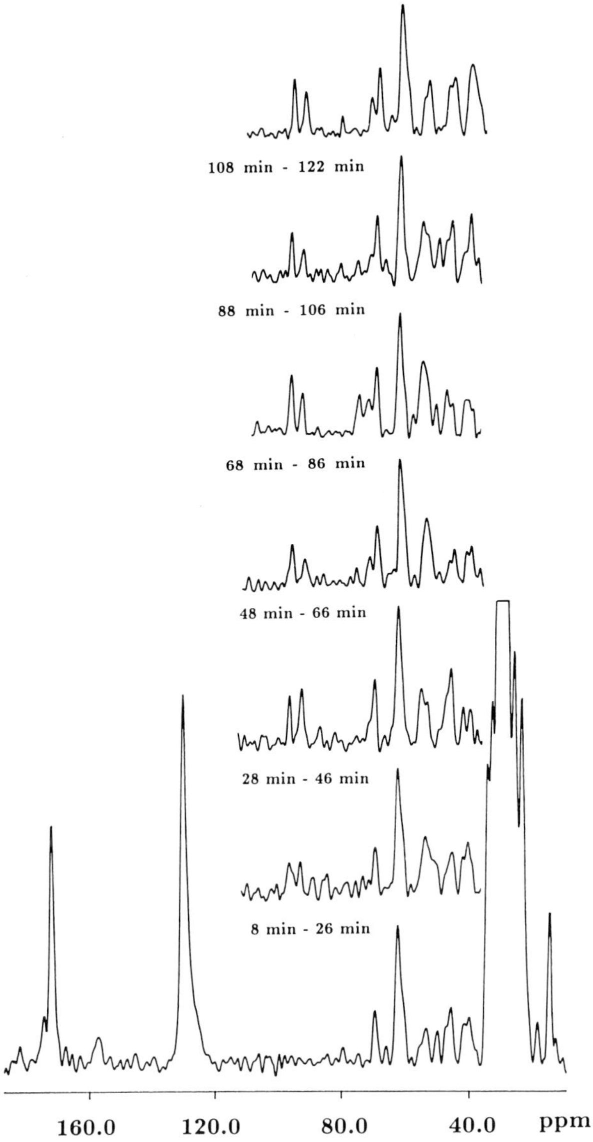


**Fig. 7.** Broadband decoupled $^{13}C$ spectra of the human head in vivo, during intravenous infusion of 20% [1 – $^{13}C$] glucose. Each spectrum represents the average of 640 acquisitions. The time scale is relative to the beginning of the glucose infusion. The resonances at 92.7 ppm correspond to $C_1$ of α-glucose, while the resonances at 96.6 ppm arise from $C_1$ of β-glucose. (Taken from Ref. [9])

dependent diabetes (NIDDM) [69]. They demonstrated that muscle glycogen synthesis is the principal pathway of glucose disposal in both normal and diabetic subjects and that defects in muscle glycogen synthesis have a dominant role in the insulin resistance that occurs in persons with NIDDM; (b) Rothman et al. observed the lactate turnover in infarcted human brain using $^{1}H$ spectroscopy after the infusion of [1-$^{13}C$] glucose. Their results suggest that elevated levels of lactate persist in the cells of stroke patients

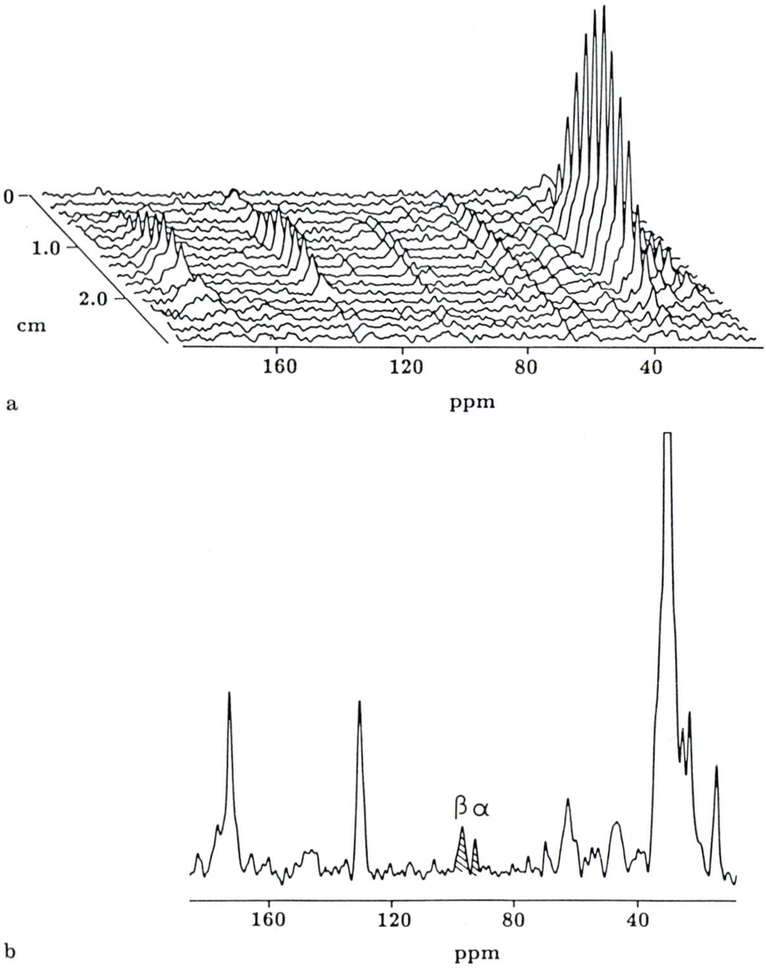


**Fig. 8 a, b.** Surface coil $^{13}C$ SI applied to the head of a normal volunteer after injection of glucose 20% labeled at the $C_1$ position. **a** Real part map. Acq. time: 9.0 min, TR = 1.05 s, 8 averages. 64 phase encoding steps (−5.0 to 5.0 mT/m). Decoupling with Waltz-8. **b** Spectrum extracted from the $^{13}C$ SI map **a**, showing the good signal-to-noise ratio of the resonances corresponding to the enriched substance, despite the low measurement time. (Taken from Ref. [9])

[64]; (c) the time course of $^{13}C$ label incorporation into the $C_2$, $C_3$, and $C_4$ resonances of glutamate/glutamine and into lactate could be recorded in the human brain with in vivo $^{13}C$ NMR [9]; (d) Beckmann et al. obtained proton decoupled $^{13}C$ NMR spectra of the human abdominal region during hyperglycemic glucose clamping using intravenous infusions of [1-$^{13}C$] glucose in normal volunteers [10]. At an enrichment level of 20% [1-$^{13}C$]

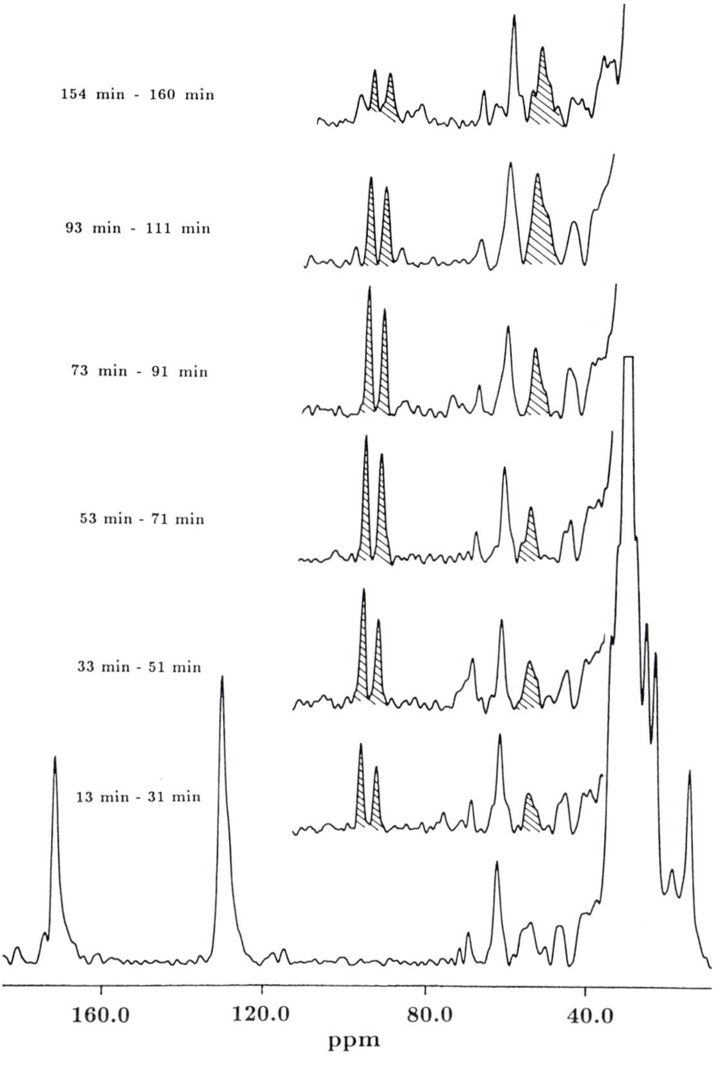


**Fig. 9.** Broadband decoupled $^{13}C$ spectra of the head of a normal volunteer, during intravenous infusion of 99% [1 − $^{13}C$] glucose. Each spectrum represents the average of 640 acquisitions. The time scale is relative to the beginning of the glucose infusion. The resonances at 92.7 ppm correspond to $C_1$ of $\alpha$-glucose, while the resonances at 96.6 ppm arise from $C_1$ of $\beta$-glucose. (Taken from Ref. [9])

glucose, the signals of α- and β-glucose at 92.7 and 96.6 ppm, respectively, were detected in the human liver already an infusion period of only 15 min. A signal arising from the $C_1$ resonance of glycogen was also detected about 20 min after the beginning of the infusion [10].

Broadband proton-decoupled $^{13}C$ spectra of the human head obtained in combination with the intravenous infusion of [1-$^{13}C$] glucose have recently been

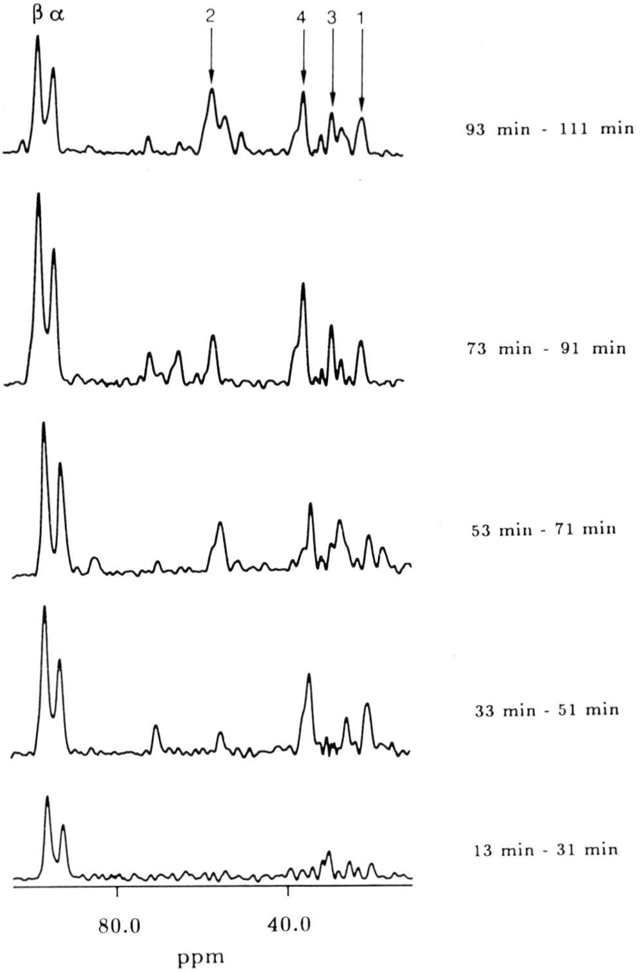


**Fig. 10.** Spectra obtained from the difference of the signals during and before the infusion, for the experiment with the 99% enriched glucose. The letters indicated in the spectra correspond to: *β*: $C_1$ of β-glucose (96.6 ppm); *α*: $C_1$ of α-glucose (92.7 ppm); *2*: $C_2$ glutamate/glutamine (55.0 ppm); *4*: $C_4$ of glutamine/glutamate (33.4 ppm); *3*: $C_3$ of glutamine/glutamate (27.8 ppm); *1*: $C_3$ of lactate (21.0 ppm). (Taken from Ref. [9])

described by Beckmann et al. [9], and are presented in the next figures. Figure 7 contains $^{13}C\{^{1}H\}$ spectra of the human head, obtained before and during the infusion of a solution enriched with 20% [1-$^{13}$C] glucose. With this concentration of enriched glucose (20%) the signals of $C_1$ of $\alpha$- and $\beta$-glucose at 92.7 ppm and 96.6 ppm, respectively, were clearly resolved.

The spatial localization of the above spectra was confirmed with $^{13}C$ spectroscopic imaging [55]. Figure 8a displays the $^{13}C$ SI spectra as a function of the slice position. The $C_1$ resonances of $\alpha$-(92.7 ppm) and $\beta$-glucose (96.6 ppm) are not observed in slices immediately adjacent to the surface coil. However, for slices deeper inside the brain (a distance of 0.8 cm to 2.0 cm) the $^{13}C$ SI spectra are characterized by significant contributions of [1-$^{13}$C] glucose (Fig. 8b). This demonstrates that the simultaneous $^{13}C$-MR examinations of serial adjacent brain regions is possible in less than 10 min [9]. This experiment points towards the direction of time-resolved, spatially-selected and chemically-specific studies in humans to obtain almost PET-like information.

The sensitivity of the measurements can be increased by using 99% enriched [1-$^{13}$C]glucose instead of only 20% enriched glucose. The spectra shown in Fig. 9 were obtained with this higher enrichment, all other experimental conditions remaining identical to those in the previous experiments [9]. As expected, the $C_1$ resonances of $\alpha$- and $\beta$-glucose are much increased in intensity compared to the experiments with 20% enrichment. These resonances remained detectable for at least 30 min after the end of the infusion. Figure 9 further demonstrates the distinct increase of the 55.0 ppm resonance after about 60 min infusion time. This resonance is attributed to $C_2$ of glutamate/glutamine [4].

Figure 10 shows the spectra obtained from the difference of the signals during and before the infusion. This comparison clearly demonstrates the appearance of resonances around 55.0 ppm and 34.4 ppm, attributed respectively to $C_2$ and $C_4$ of glutamine/glutamate [4], about 60 min after the start of the infusion. There is also an indication of additional resonances around 27.8 ppm ($C_3$ of glutamate/glutamine) and 21.0 ppm ($C_3$ of lactate). These results are comparable to analogue measurements performed by Behar et al. [11] on the rabbit brain after the scalp and the temporalis muscles had been removed.

## 8 Conclusions and Perspectives

In recent years, $^{13}C$ NMR spectroscopy has become an important tool for gaining insight into some aspects of the biochemistry of humans in vivo. Natural abundance proton decoupled $^{13}C$ spectra of the human calf, the abdominal region, and the head can now be obtained routinely in less than 30 min. In addition to lipid signals a variety of other metabolite resonances may be resolved with this temporal resolution.

The good temporal resolution has stimulated the studies with $^{13}C$-labeled substances, despite their high cost. Although $^{13}C$ NMR techniques are less sensitive than conventional radioisotope techniques and detect only metabolites which occur in the millimolar range, the $^{13}C$ approach offers some advantages:

— The kinetics of $^{13}C$ labeling in individual carbons from a variety of metabolites can be followed simultaneously and noninvasively;
— The $^{13}C$ tracers are non-radioactive;
— Different $^{13}C$ isotopomers of the same metabolite can be quantitated by the analysis of $^{13}C$–$^{13}C$ homonuclear spin coupling patterns after the administration of multilabeled substrates [21, 34, 42, 46]. The advantage of using a multilabeled substance is that certain resonances are intensified and/or different atoms in the same molecule can be visualized simultaneously. This type of information has been shown to be very useful for the interpretation of metabolic fluxes through different pathways in biological systems [24, 49].

On the other hand, the analysis of blood plasma by in vitro $^{13}C$ NMR after the infusion of $^{13}C$-labeled substances also constitutes an interesting possibility for obtaining information about the biochemistry of humans. The advantage of this kind of application is that much smaller amounts of tracer are required than for an in vivo experiment.

Further technical progress in improving the quality and the reproducibility of the $^{13}C$ spectra appears to be possible. For instance, in order to improve the spectral localization while diminishing surface signal contamination, Crowley and Ackerman proposed using electrically generated, inhomogeneous surface-spoiling magnetic field gradients [27]. Spins located within the surface-spoiling region rapidly lose phase coherence and the net magnetization produced by these spins is rendered zero. Because the magnetic field gradient exists only in the surface region, it does not affect the spin coherence of deep-lying volumes. This method could therefore provide a simple means of detecting resonances from regions farther away from the surface coil.

Operating at higher magnetic fields should be, in principle, an advantage. However, since the required total power increases with the square of the frequency range irradiated and also because of the frequency dependence of RF absorption [62], broadband decoupling should require much higher power levels at higher magnetic fields [16]. Thus, for applications on humans it remains to be verified whether the gain in sensitivity obtained at higher fields can compensate the loss of broadband decoupling possibilities at lower magnetic fields.

Another significant contribution could arise from the optimization of modulation functions to enhance the $B_1$ insensitivity and the off-resonance performance of adiabatic pulses. Important applications would be found in $^{13}C$ spectroscopy and $^{13}C$ chemical shift imaging as performed with surface coils. In vivo applications of heteronuclear proton spectral editing sequences could also benefit from the optimization of adiabatic pulses. A step in this direction was made by Garwood and Merkle, who described phantom experiments concerning the application of an adiabatic heteronuclear spectral editing sequence which is insensitive to $B_1$ inhomogeneities [33]. This sequence is based on the BIR-4 adiabatic pulse [72], which can induce uniform arbitrary flipangles despite the presence of $B_1$ inhomogeneity. The optimization of heteronuclear spectral editing could reduce the temporal resolution in experiments with enriched substrates, in order to determine the dynamics of their incorporation in a given organ.

*Acknowledgements.* The author thanks Prof. Dr. Seelig from the Biocenter for helpful comments about the manuscript. The financial support from the Swiss National Science Foundation (Grant Nr. 4.889.85.18) is gratefully acknowledged.

## 9 References

1. Alger JR, Sillerud LO, Behar KL, Gillies RJ, Shulman RG, Gordon RE, Shaw D, Hanley PE (1981) Science 214: 660
2. Alger JR, Behar KL, Rothman DL, Shulman RG (1984) J Magn Reson 56: 334
3. Aue WP, Müller S, Seelig J (1985) J Magn Reson 61: 392
4. Bárány M, Doyle DD, Graff G, Westler WM, Markley JL (1984) Magn Reson Med 1: 30
5. Bárány M, Arús C, Chang YC (1985) Magn Reson Med 2: 289
6. Beckmann N, Seelig J, Wick H (1990) Magn Reson Med 16: 150
7. Beckmann N, Müller S (1991) J Magn Reson 93: 186
8. Beckmann N, Müller S (1991) J Magn Reson 93: 299
9. Beckmann N, Turkalj I, Seelig J, Keller U (1991) Biochemistry 30: 6362
10. Beckmann N, Turkalj I, Seelig J, Keller U (1991) Proceedings of the 8th Annual Congress, Eur Soc Magn Reson Med Biol 18–21 Apr 1991, Zurich, p 133
11. Behar KL, Petroff OAC, Prichard JW, Alger JR, Shulman RG (1986) Magn Reson Med 3: 911
12. Bendall MR, Pegg DT, Doddrell DM, Field J (1981) J Am Chem Soc 103: 934
13. Bendall MR, Gordon RE (1983) J Magn Reson 53: 365
14. Bendall MR, Pegg DT (1986) J Magn Reson 67: 376
15. Bendall MR, Foxall D, Nichols BG, Schmidt JR (1986) J Magn Reson 70: 181
16. Bomsdorf H, Helzel T, Knuz D, Roeschmann P, Tschendel O, Wieland J (1988) NMR Biomed 1: 151
17. Bottomley PA (1984) US Patent 4,480,228
18. Bottomley PA, Charles HC, Roemer PB, Flamig D, Engeseth H, Edelstein WA, Mueller OM (1988) Magn Reson Med 7: 319
19. Bottomley PA (1989) Radiology 170: 1
20. Bottomley PA, Hardy CJ, Roemer PB, Mueller OM (1989) Magn Reson Med 12: 348
21. Brainard JR, Downey RS, Bier DM, London RE (1989) Anal Biochem 176: 307
22. Brown TR, Kincaid BM, Uĝurbil K (1982) Proc Natl Acad Sci USA 79: 3523
23. Canioni P, Alger JR, Shulman RG (1983) Biochemistry 22: 4974
24. Cerdan S, Seelig J (1990) Annu Rev Biophys Chem 19: 43
25. Cox SJ, Styles P (1980) J Magn Reson 40: 209
26. Cross TA, Pahl C, Oberhänsli R, Aue WP, Keller U, Seelig J (1984) Biochemistry 23: 6398
27. Crowley MG, Ackerman JJH (1985) J Magn Reson 65: 522
28. Eakin RT, Morgan LO, Gregg CT, Matwiyoff NA (1972) FEBS Lett 28: 259
29. Ernst RR Anderson WA (1966) Rev Sci Instr 37: 93
30. Ernst RR (1966) J Chem Phys 45: 3845
31. Fitzpatrick SM, Hetherington HP, Behar KL, Shulman RG (1990) J Cereb Blood Flow Metab 10: 170
32. Frahm J, Merboldt KD, Hänicke W (1987) J Magn Reson 72: 502
33. Garwood M, Merkle H (1990) Proceedings of the 9th Annual Meeting, Soc Magn Reson Med 18–24 Aug 1990, New York, p 416
34. Gopher A, Vaisman N, Mandel H, Lapidot A (1990) Proc Natl Acad Sci USA 87: 5449
35. Heerschap A, Luyten PR, van der Heyden JI, Oosterwaal LJMP, den Hollander JA (1989) NMR Biomed 2: 124

36. Hers HG, Hoof F, Barsy T (1989) in: Scriver CR, Beaudet AL, Sly WL, Valle D (eds) The metabolic basis of inherited disease, 6th edn, McGraw-Hill, New York, vol I, p 425
37. Hetherington HP, Fitzpatrick SM, Behar KL, Shulman RG (1988) Proceedings of the 7th Annual Meeting Soc Magn Reson Med 20–26 Aug 1988, San Francisco, p 752
38. Holm CH (1957) J Chem Phys 26: 707
39. Hoult DI, Richards RE (1976) J Magn Reson 24: 71
40. Jakobsen HJ, Sørensen OW, Bildsoe H (1983) J Magn Reson 51: 157
41. Jue T, Rothman DL, Tavitian DA, Shulman RG (1989) Proc Natl Acad Sci USA 86: 1439
42. Kalderon B, Korman SH, Gutman A, Lapidot A (1989) Proc Natl Acad Sci USA 86: 4690
43. Knüttel A, Kimmich R, Spohn KH (1990) J Magn Reson 86: 526
44. Knüttel A, Kimmich R, Spohn KH (1991) Magn Reson Med 17: 470
45. Knüttel A, Spohn KH, Kimmich R (1990) J Magn Reson 86: 542
46. Künnecke B, Cerdan S (1989) NMR Biomed 2(5/6): 274
47. Lauterbur PC (1957) J Chem Phys 26: 217
48. Levitt MH, Freeman R, Frenkiel T (1983) In: Waugh JS (ed) Advances of Magnetic Resonance Academic New York, vol II, p 47
49. London RE (1988) Progr NMR Spectr 20: 337
50. Luyten PR, Marien AJH, Den Hollander JA (1990) 10th European Experimental NMR Conference, Book of Abstracts, Veldhoven, The Netherlands, p 39
51. Mispelter J, Tiffon B, Quiniou E, Lhoste JM (1989) J Magn Reson 82: 622
52. Müller S, Sauter R, Weber H, Seelig J (1988) J Magn Reson 76: 155
53. Müller S (1988) Proc of the 10th Ampere Summer School and Symposium, Portoroz, p 22
54. Müller S, Hafner HP, Beckmann N (1989) NMR Biomed 2: 209
55. Müller S, Beckmann N (1989) Magn Reson Med 12: 400
56. Neurohr KJ, Barrett EJ, Shulman RG (1983) Proc Natl Acad Sci USA 80: 1603
57. Ordidge RJ, Connelly A, Lohman JAB (1986) J Magn Reson 66: 283
58. Pahl-Wostl C, Seelig J (1986) Biochemistry 25: 6799
59. Pahl-Wostl C, Seelig J (1987) Biol Chem Hoppe-Seyler 368: 205
60. Pegg DT, Doddrell DM, Bendrall MR (1982) J Chem Phys 77: 2745
61. Reo NV, Siegfried BA, Ackerman JJH (1984) J Biol Chem 259: 13664
62. Roeschmann P (1987) Med Phys 14: 922
63. Rothman DL, Behar KL, Hetherington HP, Shulman RG (1985) Proc Nat Acad Sci USA 82: 1633
64. Rothman DL, Howseman AM, Graham GD, Lantos G, Petroff OAC, Brass LM, Fayad PB, Shulman GI, Shulman RG, Prichard JW (1990) Proceedings of the 9th Annual Meeting, Soc Magn Reson Med 18–24 August 1990, New York, p 107
65. Rothman DL, Novotny EJ, Shulman GI, Howseman AM, Mason GF, Petroff OAC, Nixon T, Hanstock CC, Prichard JW, Shulman RG (1990) Proceedings of the 9th Annual Meeting, Soc Magn Reson Med 18–24 Aug 1990, New York, p 986
66. Saner M, Mckinnon G, Boesiger P (1989) Proceedings of the 8th Annual Meeting Soc Magn Reson Med 12–18 Aug 1989 Amsterdam, p 602
67. Seelig J, Burlina AP (1991) Clim Chin Acta (in press)
68. Shulman GI, Rothman DL, Smith D, Johnson CM, Blair JB, Shulman RG, Defronzo RA (1985) J Clin Invest 76: 1229
69. Shulman GI, Rothman DL, Jue T, Stein P, Defronzo RA, Shulman RG (1990) N Engl J Med 322: 223
70. Sillerud LO, Halliday KR, Griffey RH, Preiser CF, Sheppard S (1988) Magn Reson Med 8: 224
71. Silver MS, Joseph RI, Hoult DI (1985) Phys Rev A 31: 2753
72. Staewen RS, Johnson AJ, Ross BD, Parrish T, Merkle H, Garwood M (1990) Invest Radiol 25: 559
73. Stevens AN, Iles RA, Morris PG, Griffiths JR (1982) FEBS Letters 150: 489
74. Styles P, Smith MB, Briggs RW, Radda GK (1985) J Magn Reson 62: 397
75. Tiffon B, Mispelter J, Lhoste JM (1986) J Magn Reson 68: 544
76. Ugurbil K, Garwood M, Rath AR (1988) J Magn Reson 80: 448
77. US Food and Drug Administration, Rockville, MD, "Guidance for Content and Review of a Magnetic Resonance Diagnostic Device 510 (k) Application", communication to manufacturers, August 2, 1988
78. Werner B, Boesch C, Gruetter R, Martin E (1990) Proceedings of the 9th Annual Meeting, Soc Magn Reson Med 18–24 Aug 1990, New York, p 868
79. Young FE (1988) Fed Regist (FDA) 53: 7575

# Fluorine-$^{19}$F NMR Spectroscopy and Imaging In-Vivo

**M. J. W. Prior, R. J. Maxwell, and J. R. Griffiths**

CRC Biomedical Magnetic Resonance Research Group, St. George's Hospital Medical School, Division of Biochemistry, Cranmer Terrace, London SW17 ORE, UK

## Table of Contents



NMR Basic Principles and Progress, Vol. 28

The $^{19}$F nucleus is useful for studies using NMR in vivo as it has a high sensitivity and there is a low background signal from the body. The $^{19}$F nucleus has a wide range of chemical shifts which offer advantages for studies using $^{19}$F NMR spectroscopy, though they can be a problem for imaging experiments. Drugs, such as fluoropyrimidines, perfluorocarbon blood substitutes, anaesthetics and folate analogues which contain the $^{19}$F label, may be studied in animals and man. Information can be obtained by $^{19}$F NMR in vivo on the metabolism and biodistribution of these drugs. Furthermore, there are a number of fluorinated probes which may be used by $^{19}$F NMR in vivo to study the concentration of intracellular metal ions, pH, cellular hypoxia and glucose metabolism in the brain.

## List of Abbreviations

| | |
|---|---|
| APTRA | *o*-aminophenol-*N*,*N*,*O*-triacetic acid |
| BAPTA | 1,2-bis(*o*-aminophenoxy)ethane-*N*,*N*,*N'*,*N'*-tetraacetic acid |
| CB3988 | *C2*-desamino-*C2*-methyl-*N10*-propargyl-2'-trifluoromethyl-5,8-dideazafolic acid |
| CCl-103F | 3-(1,1,1,3,3,3-hexafluoroprop-2-oxy)-1-(2-nitroimidazol-1-yl)-propan-2-ol |
| $ED_{50}$ | Dose of anaesthetic effective in 50% of cases |
| Fala | fluorinated derivatives of α-methylalanine |
| F-anil | *N*,*N*-(methyl-2-carboxyisopropyl)-4-fluoroaniline |
| FBALA | α-fluoro-*β*-alanine |
| FBAPTA | 5,5'-difluoro-1,2-bis(*o*-aminophenoxy)ethane-*N*,*N*,*N'*,*N'*-tetraacetic acid |
| 5FC | 5-fluorocytosine |
| Fcryp-1 | 3,2–8,9-bis[1-fluoro-3,4-benzol]-14,15-[1-di(ethoxycarbonyl-methyl)amino-2-acetyl-4,5-benzol]-21-[3-di(ethoxycarbonyl-methyl)aminophenyl]-4,7,13,16-tetraoxa-1,10–21-triazabicyclo[8,8,5]tricosane |
| FDG | 2-fluoro-2-deoxy-D-glucose |
| FDG-6-P | 2-fluoro-2-deoxy-D-glucose-6-phosphate |
| FdUDP | 5-fluoro-2'-deoxyuridine-5'-diphosphate |
| 2'FdURD | 2'-deoxy-5-fluorouridine |
| 5'FdURD | 5'-deoxy-5-fluorouridine |
| FdUTP | 5-fluoro-2'-deoxyuridine-5'-triphosphate |
| FNUC | fluoronucleotide |
| Fquene | 8-[bis(ethoxycarbonylmethyl)amino]-6-fluoro-2-[*trans*-2[bis(ethoxycarbonylmethyl)amino]styryl]quinoline |
| Ftorafur | *R*,*S*-1-[tetrahydro-2-furyl]-5-fluorouracil |
| 5FU | 5-fluorouracil |

| | |
|---|---|
| $FUH_2$ | dihydrofluorouracil |
| FU-DNA | deoxyribonucleic acid incorporating 5FU |
| FUDP | 5-fluorouridine-5′-diphosphate |
| FUMP | 5-fluorouridine-5′-monophosphate |
| FUPA | Fluoroureidopropionic acid |
| FURD | 5-fluorouridine |
| FU-RNA | ribo nucleic acid incorporating 5FU |
| FUTP | 5-fluorouridine-5′-triphosphate |
| hplc | high pressure liquid chromatography |
| i.p. | intraperitonial |
| ISIS | image selected in vivo spectroscopy |
| i.v. | intravenous |
| PET | positron emission tomography |
| PCr | phosphocreatine |
| PFCE | perfluorocarbon emulsion |
| pKa | acid/base dissociation constant |
| Pi | inorganic phosphate |
| ppm | parts per million |
| PRPP | phosphoribosyl pyrophosphate |
| Quene 1 | 8-[bis(ethoxycarbonylmethyl)amino]-6-methoxy-2-[*trans*-2[bis(ethoxycarbonylmethyl)amino]styryl]quinoline |
| Ro-07-0741 | 3-fluoro-1-(2-nitroimidazole-1-yl)propan-2-ol |
| $T_1$ | spin-lattice relaxation time |
| $T_2$ | spin–spin relaxation time |
| TFT | 5-trifluorothymidine |

# 1 Introduction

This chapter is primarily concerned with the detection and study of fluorine-containing compounds using $^{19}$F NMR in vivo. This involves the introduction of a $^{19}$F-labelled compound into a living system as there are no $^{19}$F NMR signals normally detectable in humans or laboratory animals. Many investigations have been carried out in vitro using the $^{19}$F nucleus as a probe for chemical structures. Investigations in vitro are not discussed within this chapter, except in the few cases where they have a direct bearing on particular $^{19}$F NMR investigations in vivo.

$^{19}$F NMR has been extensively used in vivo to monitor the uptake and metabolism of drugs and, by using probe molecules, to measure a variety of parameters such as pH, hypoxia and metal ion concentration. Why is the $^{19}$F nucleus such a useful label for these applications? We can consider how well it satisfies properties that would be required of an "ideal" label and compare these with other NMR and non-NMR nuclei.

## *1.1 Sensitivity*

NMR is a relatively insensitive technique but $^{19}$F is one of the most sensitive nuclei and has 100% natural abundance (Table 1). Radioactive labels have much higher sensitivity but do not provide non-invasive chemical information and may be limited by safety considerations. The detection limits, in vivo, for compounds containing $^{31}$P and $^{1}$H are of the order of 0.5 mM and 0.1 mM, respectively, though these values may vary depending on field strength, data collection time and sample size [1]. In studies of 5FU metabolism using $^{19}$F NMR the minimum detectable concentration of $^{19}$F containing compounds was found to be of the order of 0.1 μmol g$^{-1}$ tissue (0.1 mM) in the rat [2] and 0.1 μmol g$^{-1}$ in the mouse [3]. $^{19}$F has the advantageous property (like $^{1}$H,

**Table 1.** The NMR sensitivity and natural abundance of some nuclei

| Isotope | Relative sensitivity in NMR (% of $^{1}$H sensitivity) | Natural abundance (%) |
|---|---|---|
| $^{1}$H | 100.0 | 99.98 |
| $^{19}$F | 83.3 | 100.00 |
| $^{31}$P | 6.6 | 100.00 |
| $^{13}$C | 1.59 | 1.11 |
| $^{2}$H | 0.97 | 0.02 |
| $^{3}$H | 121.0 | 0.00 |

$^{13}$C and $^{31}$P) of a spin quantum number of 1/2. These nuclei therefore have no quadrupolar moment and hence this source of line-broadening (causing signal attenuation and reduced resolution) is absent (cf $^{2}$H).

## *1.2 Non-Invasive Detection*

The main advantage of NMR is its non-invasive nature and this applies of $^{19}$F studies of fluorinated drugs. However, the administration of fluorinated probe compounds is inevitably invasive. The potential for probes to affect the parameter being measured or to cause general toxicity must be considered.

## *1.3 Background Signals*

There is negligible background signal from fluorinated compounds in the body, in contrast to the very high concentrations of water and fat protons that cause resolution and dynamic range problems in $^{1}$H MRS. The low natural abundance of $^{13}$C (Table 1) means that there is relatively little background signal so that isotopic enrichment at a specific location in a molecule can produce useful probes. Similarly, $^{19}$F may be added to a compound to enable study by NMR. It is often desirable to perform proton decoupling to simplify and enhance the signal/noise ratio of $^{13}$C spectra, which is not often necessary with $^{19}$F NMR studies in vivo. For these reasons $^{19}$F NMR experiments are often technically easier than $^{1}$H or $^{13}$C experiments and are therefore more likely to succeed.

## *1.4 Chemical Shift*

The resonant frequency of a nucleus (e.g. $^{19}$F) in a NMR experiment is proportional to the strength of the applied magnetic field. Changes of molecular structure due to the metabolism of a $^{19}$F-labelled drug which influence the distribution of electrons surrounding the $^{19}$F nucleus can be detected as a change in the chemical shift of the $^{19}$F peak. Generally, the greater the perturbation of the electrons surrounding the nucleus the greater the chemical shift difference between the parent compound and the metabolite. Thorough discussions of $^{19}$F chemical shifts are available in certain NMR texts [4] and reviews [5] which include theoretical aspects and tables of $^{19}$F chemical shifts. Here we will attempt to cover only those essentials of the topic of interest in biological applications.

$^{19}$F has a very wide chemical shift range ($>300$ ppm for organic compounds, some of which are provided for reference in Table 2). $CFCl_3$ is used as the primary chemical shift Ref. [4] but many papers, especially those reporting applications in vivo, use a variety of secondary references because of the wide range of chemical shifts and lack of a generally acceptable compound. Even for $^{19}$F applications in chemistry, external references have often been preferred to

**Table 2.** Chemical shifts of compounds studied by $^{19}F$ NMR in vivo

| Compound | Chemical shift | Source |
|---|---|---|
| $CFCl_3$ | 0.0 | 4 |
| $CF_3C_6H_5$ | −63 | 4 |
| $CF_4$ | −64 | 4 |
| $CF_3COOH$ | −77 | 4 |
| CCl-10$_3$F[a] | −73 | 91 |
| $F^-$ | −121 | 4 |
| Halothane | −158 | 73 |
| $C_6F_6$ | −163 | 4 |
| Fluoronucleotides | −166 | 101 |
| 5-Fluorouridine | −167 | 18 |
| 5-Fluorodeoxyuridine | −167 | 101 |
| Fluorocytidine | −167 | 101 |
| 5-Fluorocytosine | −169 | 50 |
| 5-Fluorouracil | −171 | 101 |
| Fluoroureidopropionic |  | 4 |
| acid | −188 | 4, 101 |
| *N*-carboxyl-fluoro-*β*- |  |  |
| alanine | −188 | 4 |
| Fluoro-*β*-alanine | −190 | 4, 101 |
| Fluorocitrate | −193 | 101 |
| 5,6-dihydrofluorouracil | −203 | 4, 101 |
| $CFH_3$ | −210 | 4 |
| Fluoroacetate | −219 | 101 |
| Ro-07-0741[a] | −234 | 91 |

[a] Structures are described in Fig. 7

internal references. Where reference compounds have been used in vivo, these are almost inevitably external. However, the usual problem of magnetic susceptibility differences between a sample and an external reference is compounded by the poor magnetic field homogeneity occurring with intact biological samples. This can lead to errors of several ppm and can therefore make external chemical references of limited value in vivo. There are relatively large solvent effects (of the order of 5 ppm) which could result in significant chemical shift differences between a drug in aqueous and lipid environments. Generally, $CF_3R$ (where R is an organic group) is shifted further upfield than $CF_2R_2$, which is further upfield than $CFR_3$. However, there may be considerable overlap between fluorine resonances from atoms attached to saturated and unsaturated carbon atoms (cf $^1H$) which may complicate peak assignments. Although the chemical shift range is large, compounds of similar structure may show only small chemical shift differences, of the same order as in $^1H$ NMR, so that minor metabolic modifications are sometimes difficult to identify in vivo.

## *1.5 Ease of Quantitation*

The quantitation of NMR spectra in vivo is not yet satisfactory. In order to obtain quantitative NMR data (i.e. the concentration of a drug in a tissue), it is necessary to compare the peak integral from the compound of interest with that from an internal or external reference of known concentration. However, the sensitivity of the radiofrequency coils used for most studies in vivo varies greatly according to position. This makes it difficult to obtain absolute quantitative data since the sensitivity of the coil to an external standard will not necessarily be the same as for the sample. The lack of background $^{19}$F signal means that there is no internal standard available. In contrast Evelhoch et al. demonstrated the use of the natural abundant deuterium signal from water in quantifying a deuterated drug by $^{2}$H NMR. The wide chemical shift range of $^{19}$F and the lack of background signals usually makes integration easier than for most other nuclei. The relative concentrations of a drug and its metabolites and the change in drug concentration with time are the parameters most frequently reported from $^{19}$F NMR studies in vivo. An additional complication is that very broad signals (as may arise when there is macromolecular binding of a drug) are very difficult to detect and integrate. Any quantitation is therefore usually of the unbound drug fraction.

### *1.6 Spatial and Temporal Resolution*

In NMR, the spatial and temporal resolution that can be achieved are limited by the sensitivity. In contrast, despite the much higher sensitivity of $^{18}$F Positron Emission Tomography (PET), its spatial resolution is limited by the uncertainty regarding the distance travelled by a positron (up to several mm) before it produces two gamma-rays, which are detected. Furthermore, a variable angle between paths of the two gamma-rays results in additional decreased resolution. The minimum volume of a cuboid selected by spectroscopic localisation techniques (e.g. ISIS) is related to the resonance frequency i.e. smaller volumes can be selected (for a given gradient strength) for nuclei having higher resonance frequencies ($^{1}$H and $^{19}$F) than for $^{31}$P, $^{13}$C or $^{2}$H.

Using $^{19}$F NMR it may be possible to determine, in vivo, not only the metabolism of a fluorinated compound but its mobility and its environment. These three features have formed the basis of many $^{19}$F NMR studies and have enabled the development of several interesting and useful probes. Many of these probes may be developed for use in the clinic or as routine research tools.

## 2 Technical Aspects

### *2.1 Localisation*

A fundamental problem in obtaining NMR spectra in vivo is to ensure that they arise from a localised region of the sample (experimental animal or human patient). A large number of techniques have been described, mostly for use in $^{1}$H or $^{31}$P experiments; the use of localisation techniques in $^{19}$F NMR studies has yet to reach its full potential. Here we include a brief introduction to the topic as it applies to $^{19}$F NMR performed in the laboratory or in the clinic, where localised $^{19}$F NMR may find its greatest application. For heterogeneous tissues and tumours, even the best current localisation methods are inadequate, but they provide data comparable to much "traditional" biochemical analysis (i.e. excision of organs followed by chemical extraction) and, of course, they have the advantage of being non-invasive.

The simplest and most common method involves the use of a surface coil [7] with a pulse-acquire sequence. The localisation achieved depends on the sensitivity profile of the surface coil; the sensitivity is maximal close to the coil and falls with increasing distance. The effective "sensitive volume" can be reduced by using a smaller coil and its position can, to some extent, be moved by the use of depth pulses [8], but there is no sharp cut-off at the edge of the sensitive volume and there is considerable variation in sensitivity within the region of interest. Surface coil spectra are prone to contamination from neighbouring tissues especially if the concentration of a given chemical species is higher in the surrounding tissues than in the tissue of interest. It is not clear to what extent $^{19}$F studies may be subject to these problems; this will depend on the fluorinated drugs used. For example, a surface coil was used to monitor the clearance of an antifolate anticancer drug from the "liver region" of mice [9]. Subsequent excision of tissues and drug assay by high pressure liquid chromatography (hplc) showed that the concentration was 10–20 fold higher in bile than in liver. Extensive protein binding of the drug was thought to result in attenuation of $^{19}$F signal from the liver so that most of the observed signal was probably from the gall bladder. In other studies it may be necessary to account for the effect of relatively high drug concentrations in fat or in the peritoneal cavity (following i.p. injection).

The ISIS localisation technique [10] uses magnetic field gradients and selective pulses to define a cuboid. This is chosen (on the basis of a previous $^{1}$H NMR image) to include only the tissue of interest. Like the simple pulse-acquire sequence, ISIS has the advantage that data is acquired directly after the final 90° pulse; there is no attenuation of signal due to spin–spin relaxation ($T_2$) within the sequence. This optimises the sensitivity and makes it easier to obtain quantitative information about drugs and their metabolites. One disadvantage is that the unwanted signal is subtracted over a cycle of 8 FIDs; there is the potential of subtraction errors if a small cuboid is selected within a large sensitive volume. Because of the large chemical shift range in $^{19}$F NMR a more serious problem is that of the chemical shift artifact. The chosen volume will be in the correct location for a $^{19}$F signal exactly on-resonance, but the voxels corresponding to off-resonance signals will be displaced in three-dimensions. For example, at 4.7 T and with a gradient strength of 20 mT/m, the centre of the ISIS voxel of 5-fluorouracil will be displaced by about 5 mm in each dimension from that of its main catabolite, fluoro-beta-alanine (19.2 ppm from 5FU). Therefore, for a 1 $cm^3$ voxel, there will be only 13% overlap between the voxels for these two metabolites. This problem is reduced by using a lower static magnetic field or by higher magnetic field gradients.

The alternative localisation techniques involve magnetic field gradients applied during a spin echo period. These allow spectra to be obtained from a single acquisition (i.e. there are no subtraction errors). However, there will be signal attenuation due to $T_2$ relaxation which may significantly influence the sensitivity and make quantitation more difficult. Scalar coupling may also cause problems. The requirement for selective pulses can result in attenuation of off-resonance signals, a potential limitation given the large chemical shift dispersion of $^{19}$F.

## 2.2 Imaging

$^1$H NMR imaging of the distribution of water (and fat) in vivo is a well established clinical diagnostic technique. Many different pulse sequences have been developed to increase image contrast in specific tissues or to overcome the problems of imaging regions of the body which move (e.g. the heart or blood flow). The contrast seen in $^1$H images depends on differences in longitudinal relaxation time ( $T_1$) and ( $T_2$) relaxation times of water in different tissues in addition to any variation in water content. Changes in these parameters in disease provide the basis for the diagnostic value of the technique. $^{19}$F images may be, in principle, obtained and manipulated in a similar fashion to gain better resolution and contrast.

The main differences in the application of imaging techniques to $^{19}$F studies result from the much lower concentration of fluorinated drugs and probes compared to tissue water, the lack of background signal and the possibility of large chemical shift effects. Signal to noise constraints have limited $^{19}$F imaging studies to compounds which can be administered in relatively high doses such as perfluorocarbon blood substitutes, anaesthetics and some anticancer drugs. Even with these drugs it is usually necessary to modify $^1$H imaging techniques in order to optimise the available signal to noise. For example, the sensitivity may be improved by the use of a high magnetic field strength, suitable choice of radiofrequency coils, short spin-echo times, and by comprorising the spatial and temporal resolution.

## 2.3 Clinical Studies

There are, at present, few examples of clinical $^{19}$F spectroscopy, and these have usually been performed with simple pulse-acquire methods. The main difficulty with fluorine spectroscopy is the necessity of introducing a $^{19}$F labelled compound, in sufficient amounts to be detected, into a patient or healthy volunteer.

Some fluorinated substances are normally administered to patients in large enough quantities to enable detection by $^{19}$F NMR in clinical systems. A good examples of this is the anti-cancer drug 5-fluorouracil (5FU), given as $1\,g\,m^{-2}$ body surface area (approximately equivalent to $14\,mg\,kg^{-1}$), which has been detected using a clinical system [11]. Since then, there have been other studies on the uptake and the fate of 5FU in patients undergoing therapy [12–14]. Studies, concerning the metabolism of 5FU in healthy volunteers, would be unethical. Jynge et al. [15], however, have detected the presence of fleroxacin (a fluoroquinolone antibiotic agent) in the liver and calf muscle of healthy human subjects, following oral administration.

Conventional studies of drug metabolism involve the examination of body fluids or extracts of biological samples. $^{19}$F NMR can also be used here, for the measurement of fluorinated drugs and their metabolites in vitro. Its main advantages are that no extraction procedures or treatments of the sample are usually required and that a range of fluorine-labelled compounds may be detected by a single experiment in blood or urine [16].

In a study on the metabolism of 5′-deoxy-5-fluorouridine (5′FdURD) in patients undergoing therapy, Malet-Martino et al. [17] used $^{19}$F NMR on samples of blood and urine to reveal the importance of catabolism and discover previously undetected metabolites of the drug. In vitro, $^{19}$F NMR has further been used to compare 5FU metabolism in patients receiving 5FU with or without methotrexate pre-treatment [18]. This study showed no effect of methotrexate pre-treatment on the catabolism of 5FU.

Clinical studies using $^{19}$F NMR, whether in vivo or in vitro, will probably become more common in the future. In vitro, it has advantages over established methods of examination of biofluids, but its value in vivo has yet to be established. Studies performed in the clinic are in some ways easier than those carried out on animals, particularly because of the larger volumes of tissue are observed in humans. Thus, using $^{19}$F NMR localisation techniques (see Sect. 2.1) to study an individual tissue of interest (e.g. liver, brain or tumour), might appear easier for studies on humans than in small animals, but, technical details can make this difficult. Furthermore, the stray field from a clinical, whole-body spectroscopy magnet is generally more expansive and intense than the stray field from the relatively small magnets used in animal research. This can make the performance of an apparently simple experiment more difficult. The use of equipment to monitor a patients vital signs and to administer drugs may be so difficult as to render certain experiments unfeasible. There are, as yet, many studies to be done using established drugs, such as 5FU (see Sect. 3.2), to determine if $^{19}$F NMR may help to predict or improve the outcome of therapy. Future in vivo $^{19}$F NMR studies could also include a number of specially created $^{19}$F-labelled compounds for the use in diagnosis of disease and measurements of metabolic parameters.

# 3 Drugs and Drug Metabolism

## 3.1 General Overview

$^{19}$F NMR has been used successfully to monitor the uptake and subsequent metabolism of a number of drugs, both in experimental animals and in man. Conventional studies of drug uptake (pharmacokinetics) are based on indirect measurements made on body fluids or excretory products; NMR permits direct studies on the drug at its site of action. Conventional metabolic studies can also be based on indirect measurements of catabolic products in urine, faeces etc., or they require sacrifice of experimental animals and assay of the drug and its products in various organs. NMR has the advantage that a complete metabolic pathway can be followed, non-invasively, in a single animal (or, in principle, in a normal volunteer or patient). As well as improving the scientific validity of the results, there is a potential for significantly reducing the number of experimental animals used in drug research—a development with both ethical and economic advantages.

### 3.1.1 Differentiation of a Drug from Its Metabolites

The wide chemical shift range of $^{19}$F means that it is usually possible to resolve and identify a number of different metabolites from the originally administered drug [19]. This is important for measurements in vivo, where poor magnetic field homogeneity tends to broaden lines, making resolution more difficult than for studies in vitro. However, the chemical shift difference between $^{19}$F signals from similar compounds may only be small (comparable to the $^{1}$H chemical shift difference). For instance, it is not possible to distinguish the various nucleotide and deoxynucleotide phosphate derivatives of 5-fluorouracil from each other in vivo.

### 3.1.2 Perturbation of the Physical and Biological Properties of Drugs

The replacement of $^{1}$H by $^{19}$F, which is the usual chemical substitution, may result in the modification of chemical properties of the final compound. For example, fluoroacetate is highly toxic compared to acetate [20]. The substitution of fluorine for hydrogen at position 5 in uracil, however, produces the toxic compound 5-fluorouracil (5FU) which is used as a chemotherapeutic agent [21]. Compounds such as this, which have been fluorinated to produce or modify an existing therapeutic action are therefore most suitable for studies in vivo. Examples of these are discussed below (see Sects. 3.2–3.5) and further consideration is given to a group of compounds developed for use as probes in vivo (see Sect. 4). It may also be possible to introduce a fluorine label on a part of molecule where it will cause minimal alteration of the properties [9]. It should be noted that $^{13}C$ and $^{2}H$ are more suitable than $^{19}F$ for this purpose but they have other disadvantages, particularly that of lower sensitivity.

### *3.1.3 Application of $^{19}F$ NMR to Studies on Drugs*

The uptake, metabolism and distribution of fluorinated drugs have been studied in animal models by $^{19}F$ NMR in vivo. [9, 19, 22–27]. Further studies using $^{19}F$ NMR have investigated metabolism in perfused mouse liver [28] and laboratory animals [29]. In the clinical setting $^{19}F$ NMR has been used in vivo to study drug metabolism in healthy volunteers [15], patients receiving anti-cancer chemotherapy [12–14] and in vitro to examine biofluids from patients treated for cancer [18, 30]. Studies using $^{19}F$ NMR have the potential to provide information from patients undergoing treatment that could help predict the outcome of therapy or provide information that could increase the effectiveness of some drugs. The technique of NMR has limitations and advantages, though, which affect the information that may be provided. The general nature of $^{19}F$ NMR was discussed earlier, now let us examine some specific details.

## *3.2 Fluoropyrimidines*

### *3.2.1 Anticancer Agents*

The principal fluoropyrimidine anticancer agent, 5-fluorouracil (5FU), was developed by Heidelberger et al. [31] in 1957. It is now used treat to breast and colon cancer [32] but also has activity against head and neck tumours [21], mammary tumours [33] and leukaemia [34]. The clinical effectiveness of 5FU is limited [35] so analogues of 5FU have been synthesised in attempts to increase the therapeutic ratio. These compounds are intermediates in the activation pathway of 5FU, for example 2′-deoxy-5-fluorouridine (2′FdURD) [36] or nucleoside derivatives with novel substitutions in place of a ribose or deoxyribose. The latter require conversion to 5FU to be effective as cytotoxic agents, and may be regarded as prodrugs of 5FU. The first of these prodrugs is *R,S*-1-[tetrahydro-2-furyl]-5-fluorouracil (Ftorafur) which requires cleavage of the furanyl ring to release 5FU and so be active [37, 38]. The second, 5′-deoxy-5-fluorouridine (5′FdURD) is also cleaved to 5FU [39] to produce cytotoxic metabolites but at a greater rate than Ftorafur [40].

### *3.2.2 The Pathways of Activation*

Activation of fluoropyrimidines occurs when they are converted to nucleotide derivatives [41] and these may be detected in vivo by $^{19}F$ NMR [19]. The activation of 5FU by enzymes of uridine triphosphate production will produce 5-Fluoro-uridine mono, di and triphosphates (FUMP, FUDP and FUTP respectively), whereas the enzymes of the thymidine synthesis and salvage pathways may activate 5FU to 2′-deoxy-5-fluorouridine mono, di and triphosphates (FdUMP, FdUDP and FdUTP respectively) [31]. The metabolism of the fluoropyrimidines included in this chapter has been summarised in Fig. 1. The fluoronucleotides, FdUMP and FUTP have been shown to have cytotoxic effects; FdUMP via inhibition of thymidylate synthase (5,10-methylenetetrahydrofolate:deoxyuridinemonophosphate C-methyltransferase) [41], and FUTP by its incorporation into RNA [42]. The relative importance of these pathways for the treatment of human tumours is not known.

Although $^{19}$F NMR can detect the activation of fluoropyrimidines, at the field strengths of most studies in vivo, it is not possible to distinguish between

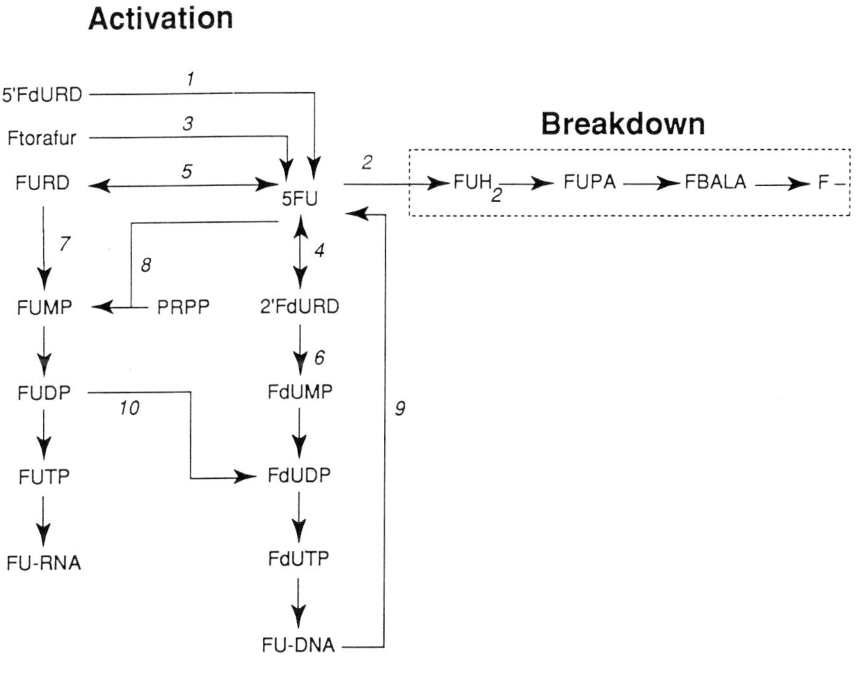


**Fig. 1.** Diagram to illustrate the metabolism of 5-fluorouracil and related compounds Abbreviations used are; 5′FdURD, 5′-deoxy-5-fluorouridine; Ftorafur, *R,S*-1-[tetrahydro-2-furyl]-5-fluorouracil; 5FU, 5-fluorouracil; $FUH_2$, dihydrofluorouracil; FUPA, fluoroureidopropionic acid; FBALA, α-fluoro-β-alanine; $F^-$; fluoride ion; FURD, 5-fluorouridine and its 5′-monophosphate, FUMP; 5′-diphosphate, FUDP; and 5′-triphosphate, FUTP; 2′FdURD, 2′-deoxy-5-fluorouridine; and its 5′-monophosphate, FdUMP; 5′-diphosphate, FUDP; and 5′-triphosphate, FdUTP; FU-DNA, deoxyribonucleic acid incorporating 5FU; FU-RNA, ribonucleic acid incorporating 5FU; PRPP phosphoribosylpyrophosphate. Enzymes are numbered as follows; *1*, pyrimidine nucleoside phosphorylase; *2*, dihydrouracil dehydrogenase; *3*, soluble and microsomal oxidative enzymes; *4*, thymidine phosphorylase; *5*, uridine phosphorylase; *6*, thymidine kinase; *7*, uridine kinase; *8*, pyrimidine phosphoribosyl transferase; *9*, uracil DNA glycosidase; *10*, ribonucleoside diphosphate reductase

FdUMP and FUTP [43]. The abundance of ribonucleotides in the cell is generally greater than that of the deoxyribonucleotides [44]. This means that the majority of fluoronucleotide (FNUC), in vivo, is most likely to consists of FUMP, FUDP and FUTP. It has been noted that 50–75% of the FNUC detectable by NMR in vivo, was probably FUTP [2, 45, 46]. Therefore, $^{19}F$ NMR probably detects changes in the ribonucleotide derivatives of fluoropyrimidines only, though NMR-visible FNUC could reflect changes in FdUMP.

The doses of 5FU required to produce NMR detectable levels of FNUC vary according to which tumour is used. Doses of 180 mg kg$^{-1}$ (i.v. bolus) were used in the Lewis Lung tumour [19] or 150 mg kg$^{-1}$ (i.p. bolus) in the mammary carcinoma [29]. However, doses lower than these may produce visible levels of FNUC in NMR spectra of experimental tumours; 120 mg kg$^{-1}$ and 50 mg kg$^{-1}$ produce NMR-detectable levels of FNUC in the Walker carcinosarcoma [2]. FNUC has been detected in human tumours in patients given 5FU by intra-arterial infusion [14]. Further studies using 5FU in experimental tumours have observed an increase in NMR visible levels of FNUC using methotrexate in the Walker carcinosarcoma [2] and choline orotate in the mouse M5076 tumour [3, 29].

The detection of the activation products of the fluoropyrimidines by $^{19}F$ NMR may be useful for the design of drug regimes which enhance activation or decrease side-effects. In NMR studies on combination chemotherapy it has been shown that FNUC levels can predict tumour response to 5FU in the Walker Carcinosarcoma [2] which is likely to be mediated via FUTP. It remains to be shown whether or not FNUC formation can predict tumour sensitivity to 5FU or its derivatives in patients.

### *3.2.3 The Pathway of Breakdown*

Although $^{19}F$ NMR may allow the detection of the activation of 5FU to FNUC in the tumour, there is only one example of this in the clinic to date [14]. However, in studies of patients undergoing chemotherapy, all have detected the breakdown of 5FU [11–14]. Studies of the catabolism of 5FU may reveal the rate of uptake and the half-life in the liver, they may indicate the extent and rate of catabolism and may allow the comparison between different drugs and different tissues.

Clearance of 5FU is very rapid and occurs mostly in the liver [47]. The route of catabolism of 5FU (see Fig. 1) is thought to be analogous to the breakdown of thymidine [48]; the ultimate step forms $\alpha$-fluoro-$\beta$-alanine (FBALA), which may liberate toxic $F^-$ ions [18]. Catabolism of 5FU occurs in most tissues, but the pathway is thought to be reduced in tumours compared with normal tissue [31] which may account for some chemotherapeutic selectivity.

Signals corresponding to 5FU, fluoroureidopropionic acid (FUPA, wrongly assigned as dihydrofluorouracil) and FBALA were first observed by $^{19}F$ NMR spectroscopy in mouse liver [19] and were subsequently identified in mouse liver and implanted tumours [3], rat liver [26] and rat tumours [2]. $^{19}$F NMR studies that have followed the breakdown of 5FU in mouse and rat liver observe signals from 5FU, FUPA and FBALA in various proportions [18, 19, 26], though in rabbit and human liver signals from only 5FU and FBALA are detected [49]. The reported extent of 5FU catabolism in the tumour appears to be less than the metabolism in the liver [18, 19, 26, 43].

Measurements of the kinetics of 5FU have been made in experimental models and in patients. In the clinical setting the half-life of 5FU has been measured in the human liver of patients undergoing chemotherapy [11]. Studies in the rat have measured the kinetics of 5FU, 2′FdURD, 5′FdURD and ftorafur [26] and the half-life of 5FU and 5′FdURD were determined in the liver. Furthermore, comparisons could be made between the rate of metabolism of all four drugs by the observation of the rate of FBALA production. It was noted that there was a link between the general toxicity of the drugs and the rate of FBALA production, suggesting that general toxicity may be governed by the availability of 5FU to the tissues.

In vitro measurements have been made of the rate of appearance of FBALA in the plasma and urine of patients receiving 5FU [18]. $^{19}$F NMR has an advantage over more conventional means (e.g. hplc) for the measurement of 5FU and its metabolites in biofluids as there is no need for preparation of the sample. It was shown that increasing the dose of 5FU did not alter the amount of catabolism and that the catabolic pathway was, therefore, saturated. Comparisons of the kinetics from different doses of 5FU have not been made in whole body systems. Observation of the rate of metabolism of 5FU in individual patients may be a useful means for limiting general toxicity, for example the infusion rate of 5FU could be controlled so as to prevent saturation of the catabolic pathway. Such whole body studies could be supported by the use of $^{19}$F NMR in vitro. Furthermore, studies on the catabolism of 5FU are a model for the study of the kinetics of other drugs.

### *3.2.4 Anti-Fungal and Antiviral Agents*

5-Fluorocytosine (flucytosine or 5FC) is used as a treatment for fungal infections [49]. The mechanism of action involves the disruption of RNA synthesis when susceptible fungi convert 5FC into 5FU using the enzyme cytosine deaminase, which is not present in human cells [51]. The metabolism of 5FC has been studied with $^{19}$F NMR in two Aspergillus species (*Aspergillus fumigatus* and *A. niger*). 5FC was found to be deaminated to 5FU and then transformed into fluoronucleotides; the catabolite $\alpha$-fluoro-$\beta$-alanine was also detected in *A.fumigatus*. Neither 5-fluoroorotic acid nor FdUMP were detected in perchloric acid extracts after any incubation with 5FC.

A further development of fluoropyrimidines resulted in the production of trifluoromethyl deoxyuridine (trifluorothymidine or TFT) which has poor activity against tumours, but is active against herpes virus [31]. This compound can be detected more easily than 5FU as it has three equivalent fluorine atoms which give three-fold higher signal for each mole of TFT.

TFT has been studied [52] as a probe for herpes infection in mice. It was found that an i.v. injection of TFT resulted in significantly higher concentrations of the drug in the livers of mice infected with herpes simplex virus than in the livers of uninfected mice, mice infected with murine hepatitis virus or mice with hepatitis from carbon tetrachloride treatment.

### *3.3 Perfluorocarbons*

Perfluorocarbons are fluorinated organic molecules which may be used as temporary blood supplements when formed into an emulsion since they have a higher capacity for carrying oxygen and carbon dioxide than does blood [53]. Perfluorocarbon emulsions (PFCE) may be studied both by $^{19}F$ spectroscopy, to measure tissue perfusion [54] and imaging to measure $pO_2$ in the blood vessels of tissues perfused with blood containing PFCE, and uptake of PFCE into different tissues [55–58].

Authier [54] followed reactive hyperaemia by using NMR spectroscopy to measure total fluorine signal from PFCE in a rat leg held within a solenoid coil. This use of $^{19}F$ NMR is a simple, though, effective way of performing a study on perfusion within a limb. One application might be to compare blood perfusion between a normal limb and one with compromised circulation. As with the majority of NMR studies in vivo, the determination of absolute quantities may be impossible. Furthermore, regional perfusion (within tissues of the limb) may not be determined using this technique, though it may be possible to obtain this information by the use of image guided localised spectroscopy or NMR imaging techniques for the measurement of flow. It may also be possible to measure oxygenation by the exploitation of the relationship between the spin-lattice relaxation time, $T_1$, of the perfluorocarbons and their oxygenation status. This technique has been implemented in imaging studies performed by Eidelberg et al. [59] and has wide range of potential applications which are discussed below.

The method of measurement of $pO_2$ relies on the relationship between oxygenation of PFCE and the value of $T_1$ for the $^{19}F$ nuclei in the compound. Eidelberg et al. [59] demonstrated a linear relationship between oxygen tension, from 20–700 Torr, and the value of $T_1$ from the $^{19}F$ NMR signals of perfluorotributylamine. Furthermore, their calibration curve used for the calculation of tissue oxygenation demonstrates the oxygen carrying capacity of the perfluorocarbons which is greater than the response range of haemoglobin, (0–50 Torr). The oxygenation of the blood within the brain was measured [59] using $T_1$ values derived from partial saturation spin-echo images. The values of blood oxygenation reported were higher than those of tissue oxygenation gained by invasive measurements on brain tissue, although the variation of oxygenation between different regions of the brain was in agreement.

Oxygenation can also be measured from $^{19}$F PFCE images by calculation of $T_1$ values from image data obtained from localised spectroscopy. A pulse sequence must, therefore, be used which weights the image according to the $T_1$ relaxation times of the sample. Measurement of oxygenation status is an interesting application of $^{19}$F NMR imaging and spectroscopy of PFCE and may have importance for the study of tumours in vivo. The oxygenation of the tumour is an important determinant of the effectiveness of radiotherapy [60].

$^{19}$F NMR imaging of PFCE has also been applied to a number of other systems. The uptake of PFCE by white blood cells has been exploited [57] to image the reticuloendothelial system of normal animals. Similarly, because of the presence of large numbers of phagocytes, it has been possible to image tumours [61].

Ths use of $^{19}$F NMR imaging of tumour tissues may prove to be an important diagnostic application of PFCE. This technique offers a method for the discovery of tumours and for determination of the size and status. Shimizu et al. [55] have used antibodies, conjugated to PFCE, to increase the $^{19}$F-labelling of tumour tissue. This was successful, but images of mice failed to reveal PFCE in the tumour in situ. However, Shimizu et al. [55] used less PFCE than other in studies performed elsewhere [57, 61]. Histology and images of excised tumour [55] showed that the antibody conjugated PFCE was mostly contained within the non-necrotic regions of the tumour. This result may warrant the development of more effective labelling of antibodies with $^{19}$F to increase the signal-to-noise ratio of the images. This may be achieved by the conjugation of antibodies with molecules that contain a large number of equivalent $^{19}$F atoms (e.g. a compound containing several —$CF_3$ groups).

The use of PFCE in tumour studies requires further work. It remains to be shown whether antibody labelling can increase PFCE concentrations in tumour tissues or whether its action is to increase PFCE label in phagocytes. This difference may not be important for tumour detection by $^{19}$F NMR imaging but will affect the interpretation of $T_1$ measurements used to determine tumour oxygenation. Bornert et al. [56] have used $^{19}$F NMR imaging to observe vasculature in isolated organs. The application of $^{19}$F imaging to studies of vasculature in vivo, may be another useful method in the clinic.

## *3.4 Anaesthetics*

The inhaled general anaesthetics are soluble in cell membranes and their action is well known to be proportional to their oil\gas partition coefficient [62]. These anaesthetics (see Fig. 2) include a number which are fluorinated such as halothane, methoxyflurane and isoflurane [63]. $^{19}$F NMR studies of fluorinated anaesthetics have been carried out, in brain [63–65], liver [24] and tumour [6, 67].

In $^{19}$F NMR studies of anaesthetic uptake into brain [63], halothane, methoxyflurane and isoflurane were detected by the use of a simple surface coil. These anaesthetics were detectable, with good signal to noise within a few minutes

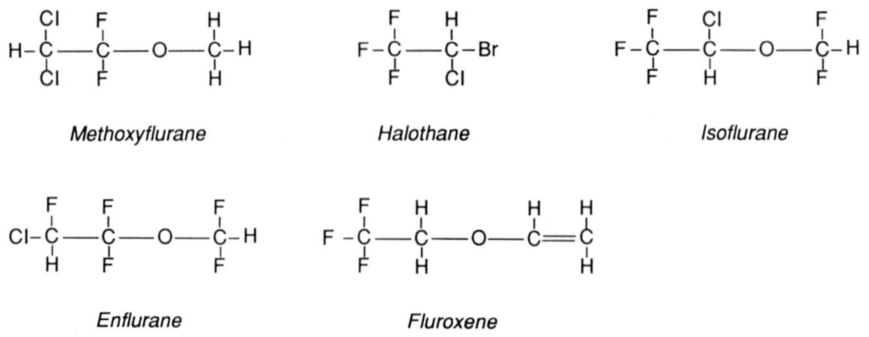

**Fig. 2.** The structure of some fluorinated volatile anaesthetics

and had a detection threshold of 0.1 to 0.5 mM [63], similar to the limit of detection reported for the fluoropyrimidines [2, 3, 19]. $^{19}$F NMR signals were detected for up to 98 hour after the anaesthetic administration was discontinued which was surprising as halothane was expected to have be eliminated after a few hours when the effects of anaesthesia have disappeared. This result was clarified, to some extent by later work [23] which revealed the presence of a less volatile, less lipid soluble metabolite of halothane which probably contained the $CF_3$ group of halothane.

Further studies [68] have used $^{19}$F NMR to examine the effect of ventilation on cerebral concentrations of fluorinated anaesthetics. It is known that mechanical ventilation and the concentration of anaesthetic can increase the alveolar concentrations of inhaled anaesthetics [68]. Furthermore, it is assumed that the concentration of inhaled anaesthetic is related to the concentration in the alveolus [69]. In $^{19}$F NMR experiments, with spontaneously breathing animals, it was demonstrated that the cerebral concentrations did not increase proportionally to inspired concentrations of halothane that exceeded 3%. It was also shown that brain concentrations were proportional to blood concentrations in mechanically ventilated animals and that these two were proportional to halothane concentration in inspired gas [68]. These results show that the inhibitory effects of halothane on ventilation limit the concentration of the anaesthetic in the brain.

Work performed by Wyrwicz [23, 70] showed that the elimination of isoflurane from brain is faster than halothane. Both Wyrwicz [23] and Rudin and Sauter [71] detected a two phased decay of halothane signals from the brain. The initial rate of decay of the $^{19}$F NMR signal from halothane was used by Rudin and Sauter [71] to measure cerebral blood flow to gain measurements in good agreement with those obtained by other methods.

The use of fluorinated anaesthetics in the clinic has been limited due to the toxicity of their metabolites [72]. The metabolism of methoxyflurane is known to produce $F^-$ ion, which may cause renal toxicity [72] and the metabolism of halothane can release trifluoroacetate [24]. The metabolism of the fluorinated anaesthetics, halothane and methoxyflurane, has been studied in liver [24, 72] using $^{19}$F NMR.

Studies on halothane [24] were performed with or without phenobarbitone induction of the cytochrome P-450 enzyme system prior to examination with $^{19}$F NMR. The metabolism of halothane in the liver is via the cytochrome P-450 enzyme system and subsequent to enzyme induction the half-life of halothane in the liver, as measured by $^{19}$F NMR, was reduced. Results of experiments with methoxyflurane showed that its elimination is a first order process [72] but that the rate was dependent upon the period of exposure to the anaesthetic.

The effects of anaesthetics are probably concerned, not only with their oil\gas partition coefficient but also with their specific sites of interaction and their mechanisms of action [62]. Studies of the mechanism of action of fluorinated anaesthetics have used $^{19}$F NMR imaging in an attempt to determined the areas of brain where anaesthetic interaction occurs [64, 73]. Spin-echo pulse sequences have demonstrated images from brains of animals exposed to halothane [64, 73], though these images are far from detailed enough to determine which regions of the brain are involved with brain/anaesthetic interactions during anaesthesia. The main problem with imaging of halothane in the brain is the long value for $T_1$ (approximately 1 s [64, 73]) and the relatively short value of $T_2$ (4.7 ms [73]; 3.5 ms [64]). This means that a long period has to be left between pulses to enable the nuclei to relax through $T_1$. The short value of $T_2$ will mean there are signal losses during the echo period of the spin–echo pulse sequence used to produce the image.

**Table 3.** Spin–spin relaxation times ($T_2$) for halothane in rabbit brain and surrounding tissue recorded at 2 *T*

| Tissue | $T_2$(ms) |
|---|---|
| Brain | 4.7 [73] |
| | 3.5 [64] |
| Muscle | 110 [73] |
| | 63.3 [64] |
| Fat | 250 [73] |

**Table 4.** Spin–spin relaxation times ($T_2$) of fluorinated anaesthetics in rat tissue recorded at 7.4T [65]

| Volatile agent | $ED_{50}$ (atm.) | $T_2$ in brain at 10 °C (ms) | $T_2$ in fat at 10 °C (ms) |
|---|---|---|---|
| Methoxyflurane | 0.0046 | 0.62 | 216 |
| Isoflurane | 0.0162 | 2.39 | 281 |
| Fluroxene | 0.0345 | 4.30 | 396 |
| Hexafluoroethane | 17 (approx.) | 18.5 | 228 |

The distribution of values of $T_2$ for halothane is not continuous throughout the head; brain has a different $T_2$ compared to surrounding tissues. Wyrwicz and Conboy [73] and Chew [64] report similar values for the $T_2$ of halothane in the brain, though different values of $T_2$ for other tissues (Table 3). Evers et al. [65] however, used $^{19}$F NMR to study the spin–spin relaxation times for a series of fluorinated anaesthetics and hexafluoroethane, not an anaesthetic, in excised brain to reveal a correlation between the anaesthetic potency ($ED_{50}$) and the value of $T_2$ (see Table 4). The lipid solubility of fluorinated anaesthetics means that they are preferentially partitioned into cell membranes and other hydrophobic environments. This feature means that they, or related compounds, may make useful probes of membrane structure or of other hydrophobic regions of tissues.

Burt et al. [66] have used halothane as a probe of membrane structure in tumours. Studies on a rat renal adenocarcinoma revealed that the tumour contained a broader resonance, possibly containing two overlapping peaks compared to normal kidney, muscle and tissue surrounding the tumour [66]. Given that this tumour is known to have an altered lipid metabolism and membrane composition the $^{19}$F NMR peak broadening may be related to an altered membrane structure in these tumour cells.

Further work by Moore and Roberts [67] studied the difference between normal kidney, renal tumour and tumour treated with hyperthermia [60]. Results of tissue extracts showed that tumour had more lipid per gram tissue than normal kidney and that the composition of the lipid was different. Examination of extracts by $^{31}$P NMR revealed that hyperthermia treatment altered the tumour lipid composition to more closely resemble that of normal tissue. Furthermore, the $^{19}$F NMR spectrum of halothane in treated tumour lipid was similar to that of halothane in normal kidney lipid.

There is further work to carried out on the interaction of halothane with membranes using both spectroscopy and imaging. Fluorinated anaesthetics or lipid soluble fluorinated gases may yet prove to be valuable probes of membrane structures or markers of altered membrane compositions. There are, at present, no non-invasive alternatives to these compounds for studying hydrophobic regions of the cell.

## 3.5 Folate Analogues

The pharmacokinetics of an antifolate anticancer agent, $C_2$-desamino-$C_2$-methyl-$N_{10}$-propargyl-2′-trifluomethyl-5.8-dideazafolic acid (CB 3988), has been monitored in mice and rats [9]. In mice the drug could be detected in the upper and lower abdomen. The half-life for clearance from the upper abdomen was found to be 28 min. Analysis by high performance liquid chromatography (hplc) following removal of tissues and biofluids revealed high (20–30 mg/ml) concentrations of drug in bile and urine. It was concluded that most of the signals detected from the upper and lower abdomen were from drug in the gall bladder and bladder respectively. Studies on rats, with and without bile duct cannulation, suggested that the drug was undergoing enterohepatic recirculation.

# 4 Probes

## 4.1 General Overview

NMR can make a number of different non-invasive, measurements in vivo, on endogenous constituents of cells and tissues but some cellular phenomena cannot be measured, easily. Examples, such as measurement of the concentration of certain metal ions, the degree of hypoxia in a tumour, the metabolism of glucose in the brain and the rate of blood flow, may arise because there is no naturally occurring NMR-detectable species which may indicate the process which we wish to detect. However, the assessment of cellular pH in $^{31}$P NMR spectra by measurement of the chemical shift between the inorganic phosphate (Pi) peak and the phosphocreatine (PCr) peak is fairly well established as a routine method. Measurements of pH from $^{31}$P NMR spectra rely on the presence of PCr and Pi in the cells under scrutiny, however not all cells contain PCr and Pi concentrations are also sometimes too low. An alternative method for the measurement of metal ions, blood flow, hypoxia, metabolism and pH is to use an indicator, such as an NMR-visible probe, which contains fluorine, for the phenomena required to be detected. There are a number of chemical compounds which may be used for these tasks, though it should be noted that measurements made with these probes are not entirely non-invasive and there is scope for further research in this field in the near future.

## 4.2 The Measurement of pH

For the detection of changes in intracellular pH using $^{19}$F NMR a pH-indicator must have an ionisable group with a $pK_a$ value close to the pH to be measured; also the ionisation of this group must be closely coupled to the $^{19}$F-containing group so as to perturb its chemical shift. Furthermore, the compound needs to be loaded into cells in sufficient quantities to gain a high signal to noise from the intracellular space. There should also be no contamination of the intracellular signals from the indicator which remains outside the cells.

A number of pH indicators designed for use in $^{19}$F NMR are fluorinated derivatives of α-methylalanine [74] (Fala), *N*,*N*-(methyl-2-carboxyisopropyl)-4-fluoroaniline (F-Anil) [75], and Fquene, a pH-sensitive derivative of the fluorescent $Ca^{2+}$ indicator quin 2 [76]. The structures of F-Anil and the Fala indicators are shown in Fig. 3 and the structure of Fquene is shown in Fig. 4. To enable loading of these compounds into cells, in sufficient quantities for

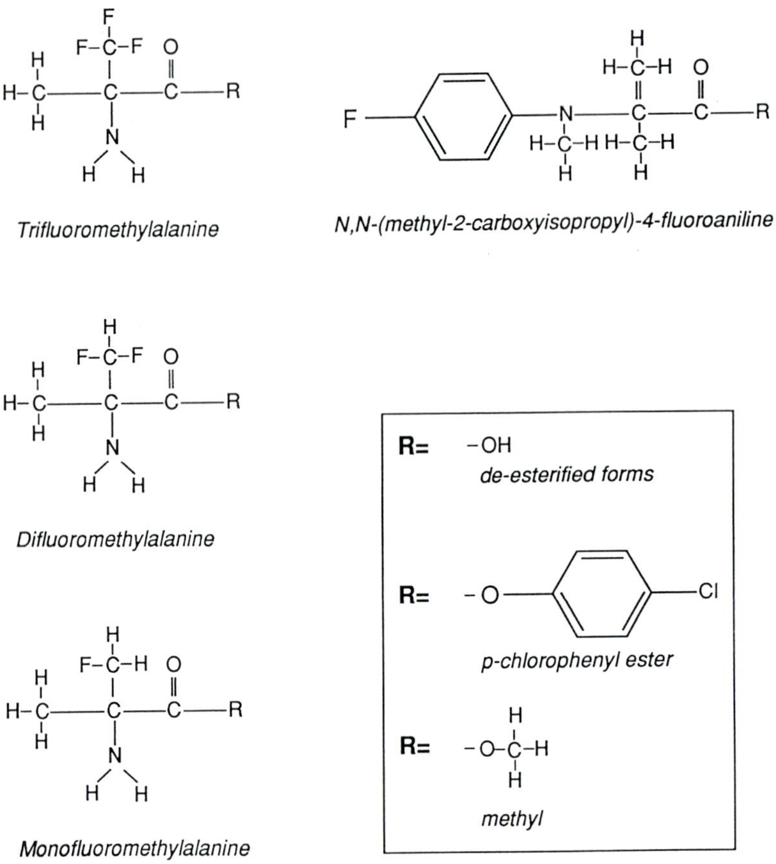


**Fig. 3.** The structure of fluorinated pH indicators based on α-methylalanine and aniline

detection by $^{19}$F NMR, they are esterified to increase their membrane solubility. Esters of the pH indicators Fala and F-anil include methyl [75] and ρ-chlorophenyl [77] derivatives (Fig. 3); Fquene [76] has been administered as the tetraacetoxymethyl derivative. Once inside the cell, esterase enzymes remove the alcohol group leaving the relatively membrane-impermeable indicator. The de-esterification process is dependent on the presence of cellular enzymes, therefore as some cells lack certain esterases (e.g. methyl esterase), the derivative of the pH indicator must be chosen for its purpose. Cell loading by esterification allows the accumulation of relatively large concentrations of intracellular indicator.

For the most sensitive measurement of in vivo pH the $pK_a$ of the indicator should be within one pH unit of the physiological value that it is required to measure. The $pK_a$ values of the fluorinated probes, shown in Table 5, imply that difluoromethylalanine, F-Anil and Fquene are the most useful for measure-

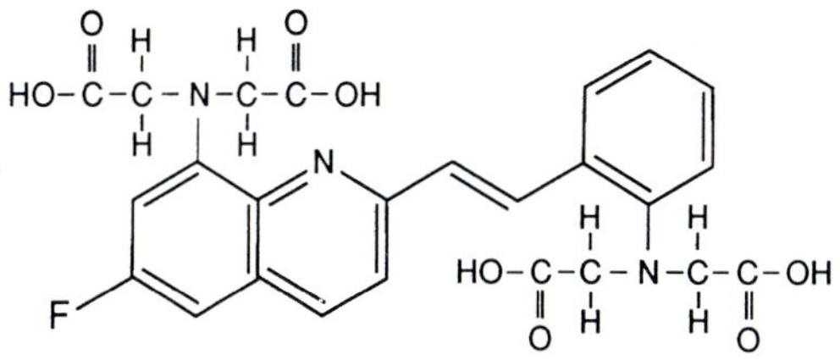

**Fig. 4.** The structure of the pH indicator 8-[bis(ethoxycarbonylmethyl) amino]-6-fluoro-2-[*trans*-2[bis(ethoxycarbonylmethyl)amino[styryl]-quinoline (Fquene)

**Table 5.** The dissociation constant, $pK_a$, for various fluorinated pH probes

| pH indicator | $pK_a$ | *Sensitivity ($\delta$ ppm) |
|---|---|---|
| α-trifluoromethylalanine | 5.9 | 1.2 |
| α-difluoromethylalanine | 7.3 | 0.9 |
| α-monofluoromethylalanine | 8.5 | 1.1 |
| F-Anil | 6.8 | 4 |
| Fquene | 6.7 | 2 |

* $\delta$ppm = change in chemical shift for a change of one pH unit away from the value of $pK_a$

ments of intracellular pH under normal conditions. However, the sensitivity of these compounds to changes in pH (see Table 5) suggests that Fquene and F-Anil would give the most accurate values.

These compounds have been used to measure pH in isolated mammalian cells and perfused heart [76], in blood cells [74, 78] in cultured mammalian cells and bacteria [77] and in cultured mammalian cells [75], however none of these pH indicators have been used in whole animals. Examples of non-invasive measurements of pH can be found in results from $^{19}$F NMR experiments with the fluoropyrimidines; fluoronucleoside and fluoronucleotide metabolites have been used to assess pH in tumours treated with their precursor, the drug 5-fluorouracil [2, 19]. Experiments with fluoropyrimidines demonstrates that it is feasible to introduce a pH sensitive compound into tumour tissue in a sufficient concentration to gain an assessment of pH, in vivo, when measurements using $^{31}$P NMR may be unfeasible, for example, when there is no PCr in the tissue. Loading of perfused organs or isolated cells with specifically designed pH indicators is possible when they are incubated with sufficient quantities of the relevant ester [75, 76]. However, the use of these detectors in whole animals is likely to be more complicated. Metabolism of the indicators is likely to occur to some extent in the liver and esterases in the blood may also destroy the esterified compound essential for cell loading. Although, toxicity studies with mice suggest that the fluoroalanines are non-toxic [79] the biodistribution of specific, pH indicators and the feasibility of their use in vivo has yet to be assessed.

## *4.3 Quantitation of Intracellular Metal Ions*

One of the most well-known NMR-indicators of intracellular metal ion concentrations is the $^{19}F$-labelled probe for intracellular $Ca^{2+}$ concentrations which was introduced by Smith et al. [80]. This is a derivative of 1,2-bis(*o*-aminophenoxy)ethane-*N*,*N*,*N*′,*N*′-tetraacetic acid (BAPTA), which has two symmetrically substituted fluorines in position 5 to give 5,5′-$F_2$-1,2-bis(*o*-aminophenoxy)ethane-*N*,*N*,*N*′,*N*′-tetraacetic acid (FBAPTA). The structure of FBAPTA is shown in Fig. 5. The role of FBAPTA as a marker of intracellular calcium in cell suspensions and tissues is complimentary to the use of fluorescent markers to make measurements in single cells. Intracellular calcium indicators that exploit fluorescence are prone to artifacts, mainly due to quenching of the fluorescence [81]. The uses of the $^{19}F$-labelled probes for measurements of intracellular ions by $^{19}F$ NMR are discussed below.

Cells can be loaded with FBAPTA by administration of the membrane permeable acetoxymethyl ester of FBAPTA. Free FBAPTA is released following hydrolysis by cellular enzymes and subsequently has a much reduced membrane permeability causing large amounts to be retained in the cell. FBAPTA has been used to measure calcium levels in perfused isolated heart [76, 82, 83] and brain slices [84–86]. However, the main problem with the assessment of intracellular $Ca^{2+}$ with FBAPTA is that it chelates a significant proportion of the $Ca^{2+}$ ions. The concentration of FBAPTA in the cell may be up to 1 mM after loading and the intracellular concentration of $Ca^{2+}$ is around 0.1 to 1 μM. Therefore, a significant proportion of the "free" intracellular $Ca^{2+}$ may be bound to FBAPTA under these conditions. In experiments with perfused heart [76] the organ had a reduced response to adrenaline when loaded with FBAPTA compared to controls. This was overcome by raising the external $Ca^{2+}$ concentration to compensate for the calcium-buffering effect of FBAPTA. Other workers [83] report that it was necessary to raise $Ca^{2+}$ concentrations in the perfusate of isolated heart, loaded with FBAPTA, to gain a full response from the organ when it was stimulated electrically. Furthermore, the use of FBAPTA to measure free intracellular $Ca^{2+}$ in brain slices [84] revealed that the buffering of intracellular $Ca^{2+}$ by the fluorinated indicator had a detrimental effect on brain tissue. However, this effect, too, was overcome by an increment of the concentration of $Ca^{2+}$ in the external perfusate.

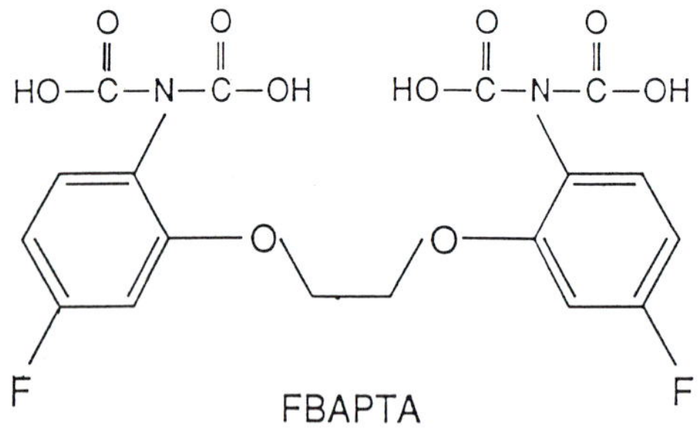


**Fig. 5.** The structure of the calcium indicator 5,5′-difluoro-1,2-bis(*o*-aminophenoxy)ethane-*N*, *N*, *N*′, *N*′-tetraacetic acid (FBAPTA)

Indicators that rely on fluorescence are prone to artifacts caused by the presence of other ions such as $Zn^{2+}$, $Fe^{2+}$, $Mg^{2+}$ and $Pb^{2+}$. These ions also bind to FBAPTA but do not interfere with the assessment of $Ca^{2+}$. This is because the different complexes between FBAPTA and intracellular ions have different chemical shifts [80] and can thus be identified, simultaneously. The measurement of $Ca^{2+}$ is gained from the relative size of the peak area for free FBAPTA compared to the peak area for FBAPTA bound to $Ca^{2+}$, allowing for the dissociation constant [76]. A similar method is employed to measure other metal ions that bind to FBAPTA.

Calcium concentrations have been measured in oesteoblasts and the effects of $Pb^{2+}$ upon the concentration of free intracellular calcium was assessed [87]. It was also possible to measure free intracellular $Pb^{2+}$. For the assessment of other metal ions it is possible to make measurements using FBAPTA and simultaneous measurements of $Ca^{2+}$ and $Zn^{2+}$ were made using FBAPTA [76], whereas $Zn^{2+}$ may interfere with the measurement of $Ca^{2+}$ using fluorescent indicators [80].

The detection of metal ions in cells and tissues may develop further as new fluorinated indicators are introduced. Smith [88] developed an indicator of intracellular $Na^+$ concentrations by means of a specific indicator, Fcryp-1, which contains some features similar to FBAPTA; the structure of Fcryp-1 is shown in Fig. 6. Fcryp-1 is loaded into cells by the administration of the acetoxymethyl ester, similarly to the administration of FBAPTA. The properties of Fcryp-1 make it valuable as an indicator of intracellular $Na^+$ in vivo. The structure of Fcryp-1 selects $Na^+$ ions by size; larger ions such as $Mg^{2+}$, $Ca^{2+}$, and $K^+$ cannot enter the binding region of Fcryp-1 to complex with it.

One further development in this field has been the production of an $^{19}$F NMR probe for intracellular $Mg^{2+}$ [89]. Based on the chelator *o*-aminophenol-*N*,*N*,*O*-triacetic acid (APTRA), the 4-fluoro, 5-fluoro, and 4-methyl-5-fluoro deriva-

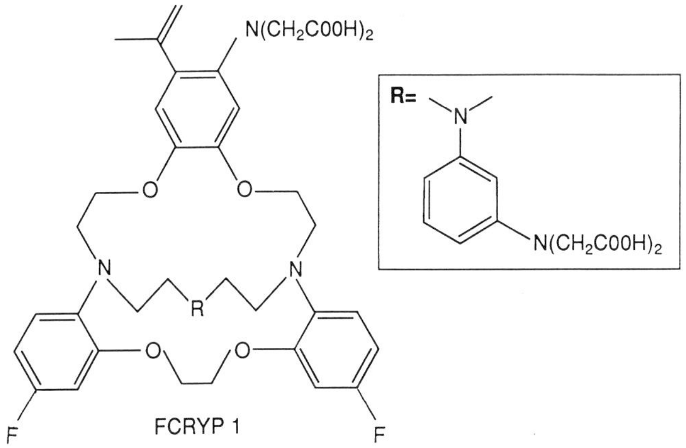


**Fig. 6.** The structure of sodium indicator 3,2–8,9-bis[1-fluoro-3,4-benzo]-14,15-[1-di(ethoxycarbonylmethyl)amino-2-acetyl-4,5-benzo]-21-[3-di(ethoxycarbonylmethyl)aminophenyl]-4,7,13,16-tetraoxa-1,10–21-triazabicyclo[8,8,5]tricosane (Fcryp-1)

tives of this compound, have dissociation constants for $Mg^{2+}$ which are higher than that for $Ca^{2+}$. However, intracellular calcium levels are generally too low to interfere with the assessment of magnesium.

$^{19}$F-labelled indicators of intracellular metal ion concentrations may be applied to many fields of research and have an important role to play in investigations of the functions of metal ions in disease and normal health. The further development of this field of research to find applications for these indicators in the areas of disease diagnosis and prognosis is a possible outcome.

## *4.4 Detection of Cellular Hypoxia*

Radiation is much less effective at killing cells under conditions of low oxygen tension than when they are fully oxygenated. Animal tumour models often contain regions which are too far from functioning blood vessels to be fully oxygenated but which nevertheless contain some viable cells. This may also be the case in human tumours and would result in resistance to radiotherapy. The traditional radiobiological measurement of this 'hypoxic fraction' involves the treatment of a large batch of tumours with various doses of radiation followed by in vitro assessment of cell viability. Such a method is clearly unsuitable for use in patients so it has been difficult to extend the studies made on animal tumour hypoxia to the clinical situation. This may partly explain the failure of treatments developed to overcome tumour hypoxia (e.g. hyperbaric oxygen and radiation sensitisers) to gain clinical acceptance. Non-invasive techniques such as gamma-scintigraphy, positron-emission tomography and $^{19}$F NMR spectroscopy have been employed to measure probe compounds which may give an indirect indication of tumour hypoxia. It is hoped that they will allow the selection of a subset of patients with hypoxic tumours who would be expected to benefit from therapeutic strategies which sensitise or even directly kill hypoxic cells.

Nitroheterocycles (in particular 2-nitroimidazoles) have been shown to undergo nitroreductive bioactivation resulting in covalent binding and selective retention in hypoxic cells. Chapman [90] administered the hexafluorinated nitroimidazole, 3-(1,1,1,3,3,3-hexafluoroprop-2-oxyl)-1-(2-nitroimidazol-1-yl)propan-2-ol (CCI-103F, see Fig. 7a) to tumour bearing mice and were able to measure $^{19}$F signals from the tumours following excision. $^{19}$F NMR has further been used to monitor the clearance of CCI-103F and a monofluorinated analogue, 3-fluoro-1-(2-nitroimidazol-l-yl)propan-2-ol (Ro-07-0741, see Fig. 7b), from tumours, liver and brain of mice in vivo [91]. It was demonstrated that there was similar initial uptake into tumours and brain at early times (1–2 hours) but selective retention of $^{19}$F signal in EMT-6 mammary tumours (and to a lesser extent KHT and RIF-1 sarcomas) at 5–6 hours. This was despite the absence of significant concentrations of the parent probe compounds in plasma and tissues at these later times when measured by hplc. These results suggested the involvement of a product of bioreductive activation and the extent of retention

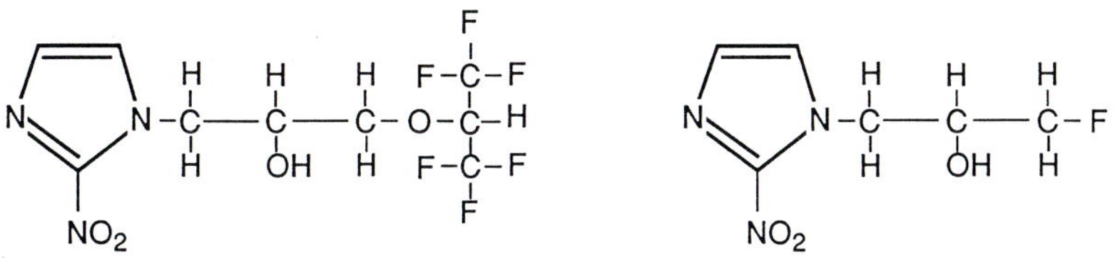


**Fig. 7 a, b.** The structure of two nitroimidazole compounds; **a** shows compound CCl-103F and **b** shows compound Ro-07-0741

in the three tumour types was consistent with the previously reported hypoxic fractions and in vivo nitroreductase activities.

## *4.5 Brain Metabolism*

The glucose analogue 2-fluoro-2-deoxy-glucose (FDG) is believed to be metabolised in the brain by the enzyme hexokinase to 2-fluoro-2-deoxy-glucose-6-phosphate (FDG-6P) in a fashion analogous to the metabolism of glucose [92]. Unlike glucose, however, FDG is not considered to be metabolised beyond this point in glycolysis with the result that FDG-6P is trapped within the cell. Original studies performed using positron emission tomography (PET) to detect the uptake of $^{18}$F-labelled FDG in brain [93, 94] suggested regional increases in glucose utilisation indicated by higher intensity of detected signals from FDG. However, as PET studies detect the radioactivity of the $^{18}$F nucleus they are unable to distinguish metabolites from the original compound and so cannot reveal as much information as $^{19}$F NMR studies using $^{19}$F-labelled FDG. Against this advantage one must set the much higher sensitivity and thus spacial and temporal resolution of the PET method. However, the sensitivity of $^{19}$F NMR can be improved by the application of $^{1}$H-decoupling [95] which improves both peak resolution and signal-to-noise. $^{19}$F NMR has been used to assess aspects of cerebral metabolism in vivo by detection of the uptake and metabolism of FDG [96–99]. Although the metabolite FDG-6P is not always detectable in vivo [99] it has been observed in vivo in brain by others using $^{19}$F NMR [96–98].

Results obtained using $^{19}$F NMR have revealed the presence of at least one more metabolite of FDG present in the brain following FDG administration [96–98]; this was provisionally identified to be 2-fluoro-2-deoxy-6-phospho-glucono-$\delta$-lactone [97]. These results support the view that the use of FDG as a probe of glucose metabolism is questionable [100]. However, the study of FDG metabolism by PET may still be considered as a metabolic probe and a diagnositic tool as the metabolism of FDG beyond FDG-6P may not occur significantly during the short period of PET studies [92]. The use of $^{19}$F NMR in such studies has the advantage that the extent of metabolism may be indicated by the intensity of metabolite peaks in the $^{19}F$ NMR spectra of the brain following FDG administration.

## 5 References


1. Tofts PS, Wray S (1988) NMR in Biomedicine 1: 1
2. McSheehy PMJ, Prior MJW, Griffiths JR (1989) Br J Cancer 60: 303
3. Hull WE, Port R, Oswald H, Kunz W (1986) Fifth Annu Meeting Soc Magn Reson Med 594
4. Gunther H (1980) In: NMR spectroscopy, John Wiley, New York, p 342
5. Emsley JW, Phillips L (1971) In: Progress in NMR spectroscopy 7, Pergamon, Great Britain, p 1
6. Evelhoch JL, McCoy Cl, Giri BP (1989) Magn Reson Med 9: 402
7. Ackerman JJ, Grove TH, Wong GG, Gadian DG, Radda GK (1980) Nature 283: 167
8. Bendall P, Lai C-M, Lauterbur PC (1980) J Mag Res 38: 343
9. Newell DR, Maxwell RJ, Bisset GMF, Jodrell DI, Griffiths JR (1990) Br J Cancer 62: 766
10. Ordidge RJ, Connelly A, Lohman JAB (1986) J Mag Res 66: 283
11. Wolf W, Albright MJ, Silver MS, Weber H, Reichardt U, Sauer R (1987) Magn Reson Imaging 5: 165
12. Glaholm J, Leach MO, Collins D, al-Jehazi B, Sharp JC, Smith TA, Adach J, Hind A, McCready VR, White H (1990) Br J Radiol 63: 547
13. Presant CA, Wolf W, Albright MJ, Servis KL, Ring R, Atkinson D, Ong RL, Wiseman C, King M, Blayney D et al. (1990) J Clin Oncol 8: 1868
14. Semmler W, Bachert-Baumann P, Guckel F, Ermark F, Schlag P, Lorenz WJ, von Kaick G (1990) Radiology 174: 141
15. Jynge P, Skjetne T, Gribbestad I, Kleinbloesem CH, Hoogkamer HF, Antonsen O, Krane J, Bakoy OE, Furuheim KM, Nilsen OG (1990) Clin Pharmacol Ther 48: 481
16. Malet-Martino MC, Martino R, Lopez A, Beteille J-P, Bon M, Bernadou J, Armand J-P (1985) Biochem Pharmacol 34: 429
17. Malet-Martino MC, Faure F, Vialaneix J-P, Palevody C, Hollande E, Martino R (1986) Cancer Chemother Pharmacol 18: 5
18. Hull WE, Port RE, Herrmann R, Britsch B, Kunz W (1988) Cancer Res 48: 1680
19. Stevens AN, Morris PG, Iles RA, Sheldon PW, Griffiths JR (1984) Br J Cancer 50: 113
20. Goldstein A, Aronow L, Sumner SM (1974) In: Principles of Drug Action: The Basis of Pharmacology, John Wiley, New York, p 357
21. Peters GJ, Braakhuis BMJ, de Bruijn EA, Laurensse EJ, van Walsum M, Pinedo HM (1989) Br J Cancer 59: 327
22. Joseph A, Davenport C, Kwock L, Burt CT, London RE (1987) Magn Reson Med 4: 137
23. Wyrwicz AM, Conboy CB, Ryback KR, Nichols BG, Eisele P (1987) Biochim Biophys Acta 927: 86
24. Selinsky BS, Thompson M, London RE (1987) Biochem Pharmacol 36: 413
25. Arndt DC, Ratner AV, Faull KF, Barchas JD, Young SW (1988) Psychiatry Res 25: 73
26. Prior MJW, Maxwell RJ, Griffiths JR (1990) Biochem. Pharmacol 39: 857
27. Maxwell RJ, Frienkiel TA, Newell DR, Bauer C, Griffiths JR (1991) Magn Reson Med 17: 189
28. Cabanac S, Malet-Martino MC, Bon M, Martino R, Nedelec JF, Dimicole JL (1988) NMR in Biomedicine 1: 113
29. Koutcher JA, Barnett DC, Martin DS, Stolfi R, Sawyer R, Cowburn D (1987) Sixth Annu Meeting Soc Magn Reson Med 108
30. Malet-Martino MC, Armand J-P, Lopez A, Bernadou J, Beteille J-P, Bon M, Martino R (1986b) Cancer Res 46: 2105
31. Heidelberger C, Danenberg PV, Moran RG (1983) Adv Enzymol 54: 5431
32. Cunningham D (1988) Br J Cancer 58: 695–699
33. Fisher B, Carbone P (1982) In: Holland JF, Frei E (Eds) Cancer Medicine, Lea and Febiger, Philadelphia, p 2025
34. Mini E, Mazzei T, Coronnello M, Criscuoli L, Gualtieri M, Periti P (1984) Chemioterapia 3: 343

35. MacKintosh J, Tattersall MHN (1987) Ann Acad Med 16: 444
36. Pinedo HM, Peters GFJ (1988) J Clin Oncol 6: 1653
37. El Sayed Y-M, Sadee W (1982) Biochem Pharmcol 31: 3006
38. El Sayed Y-M, Sadee W (1983) Cancer Research 43: 4039
39. Sommadossi J-P, Aubert C, Cano J-P, Gouveia J, Ribaud P, Mathe G (1983) Cancer Res 43: 930
40. Au JL-S, Sadee W (1980) Cancer Res 40: 2814
41. Cohen SS, Flaks JG, Barner HD, Loeb MR, Lichtenstein J (1958) Proc Nat Acad Sci USA 44: 1004
42. Chaudhuri NK, Montag BJ, Heidelberger C (1958) Cancer Res 18: 318
43. McSheehy PMJ, Griffiths JR (1989) NMR in Biomedicine 2: 133
44. Tanaka K, Yoshioka A, Tanaka S, Wataya Y (1984) Anal Biochem 139: 35
45. Cadman E, Heimer R, Benz C (1981) J Biol Chem 256: 1695
46. Benz C, Cadman E (1981) Cancer Res 41: 994
47. Warren BS, La Creta FP, Kornhauser DM, Williams WM (1987) Cancer Res 47: 5261
48. Mukherjee KL, Boohar J, Wentland D, Ansfield FJ, Heidelberger C (1963) Cancer Res 23: 49
49. Wolf W, Servis KL, El-Tahtawy A (1987b) Sixth Annu Meeting Soc Magn Reson Med 587
50. Vialeneix JP, Chouini N, Malet-Martino MC, Martino R, Michel G, Lepargneur JP (1987) Antimicrob Chemother 30: 756
51. Chouini-Lalanne N, Malet-Martino MC, Martino R, Michel G (1989) Antimicrob Agents Chemother 33: 1939
52. Rand KH, Raad I, el Koussi A, Houck HJ, Brey W, Rocca J, Loftsson T, Bodor N (1987) J Virol Methods 18: 257
53. Clark LC, Ackerman JL, Thomas SR, Millard RW (1984) Second Annu Meeting Soc Magn Reson Med 135
54. Authier B (1988) Magn Reson Med 8: 80
55. Shimizu M, Kobayashi T, Morimoto H, Matsuura N, Shimano T, Nomura N, Itoh S, Yamazaki M, Iriguchi N, Yamamoto T, Yamai S, Furuta T, Maki T, Mori T (1987) Mag Reson Med 5: 290
56. Bornert P, Dreher W, Schneider W (1989) NMR Biomed 2: :278
57. Ratner AV, Hurd R, Muller HH, Bradley-Simpson B, Pitts W, Shibata D, Sotak C, Young SW (1987) Magn Reson Med 5: 548
58. Eidelberg D, Johnson G, Tofts PS, Dobbin J, Crockard HA, Plummer D (1988) J Cerb Blood Flow Metab 8: 276
59. Eidelberg D, Johnson G, Barnes, D, Tofts PS, Delpy D, Plummer D, McDonald WI (1988) Magn Reson Med 6: 344
60. Workman P (1985) In: Farmer PB, Walker JM (eds) The molecular basis of cancer, Croom Helm, London, p 237
61. Longmaid HE, Adams DF, Neirinckx RD, Harrison CG, Brunner P, Seltzer SE, Davis MA, Neuringer L, Geyer RP (1984) Magn Reson Med 1: 197
62. Willatts SM (1990) In: Aitkenhead AR, Smith G (eds) Textbook of Anaesthesia, Churchill Livingstone, Edingburgh, p 77
63. Wyrwicz AM, Pszenny MH, Schofield JC, Tillman PC, Gordon RE, Martin PA (1983) Science 222: 428
64. Chew WM, Moseley ME, Mills PA, Sessler D, Gonzalez-Mendez R, James TL, Litt L (1987) Magn Reson Imaging 5: 51
65. Evers AS, Haycock JC, d-Avignon DA (1988) Biochem Biophys Res Commun 151: 1039
66. Burt CT, Moore RR, Roberts MF, Brady TJ (1984) Biochim Biophys Acta 805: 375
67. Moore RR, Roberts MF (1985) Biochim Biophys Acta 844: 346
68. Lockhart SH, Cohen Y, Yasuda N, Kim F, Litt L, Eger EI, Chang LH, James T (1990) Anesthesiology 73: 455
69. Nimmo WS (1990) In: Aitkenhaed AR, Smith G (eds) Textbook of Anaesthesia, Churchill-Livingstone, Edingburgh, p 139
70. Wyrwicz AM, Conboy CB, Nichols BG, Ryback KR, Eisele P (1987) Biochim Biophys Acta 929: 271
71. Rudin M, Sauter A (1989) NMR Biomed 2: 98
72. Selinsky BS, Perlman ME, London RE (1988) Mol Pharmacol 33: 559
73. Wyrwicz AM, Conboy CB (1989) Magn Reson Med 9: 21976
74. Taylor J, Deutsch C (1983) Biophys J 43: 261
75. Deutsch CJ, Taylor JS (1989) Biophys J 55: 799
76. Metcalfe JC, Hesketh TR, Smith GA (1985) Cell Calcium 6: 183

77. Taylor J, Deutsch C (1988) Biophys J 53: 227
78. Deutsch C, Taylor JS, Price M (1984) J Cell Biol 98: 885
79. Deutsch CJ, Taylor JS (1987) Ann NY Acad Sci 508: 33
80. Smith GA, Hesketh TR, Metcalfe JC, Feeney J, Morris PG (1983) Proc Natl Acad Sci 80: 7178
81. Arslan P, Di Virgilio F, Tsien RY, Pozzan T (1985) J Biol Chem 260: 2719
82. Steenbergen C, Murphy E, Levy L, London RE (1987) Circ Res 60: 700
83. Marban E, Kitakaze M, Chacko VP, Pike MM (1987) Circ Res 63: 673
84. Bachelard HS, Badar-Goffer RS, Brooks KJ, Dolin SJ, Morris PG (1988) J Neurochem 51: 1311
85. Badar-Goffer RS, Ben-Yoseph O, Dolin SJ, Morris PG, Smith GA, Bachelard HS (1990) J Neurochem 55: 878
86. Ben-Yoseph O, Bachelard HS, Badar-Goffer RS, Dolin SJ, Morris PG (1990) J Neurochem 55: 1446
87. Schanne FA, Dowd TL, Gupta RK, Rosen JF (1990) Environ Health Perspect 84: 99
88. Smith GA, Morris PG, Hesketh TR, Metcalfe JC (1986) Biochim Biophys Acta 889: 72
89. Levy LA, Murphy E, Raju B, London RE (1988) Biochemistry 27: 4041
90. Chapman JD (1984) Cancer 54: 2441
91. Maxwell RJ, Workman P, Griffiths JR (1989) Int J Radiat Oncol Biol Phys 16: 925
92. Lear JL (1990) Radiology 174: 328
93. Revich M, Kuhl D, Greenberg J, Phelps ME, Ido T, Cassella V, Fowler J, Alavia A, Sokoloff L (1979) Circ Res 44: 127
94. Pehlps ME, Huang SC, Hoffman EJ, Selin CS, Sokoloff L, Kuhl DE (1979) Ann Neurol 6: 372
95. Berkowitz BA, Ackerman JJH (1987) Biophys J 151: 681
96. Nakada T, Kwee IL (1986) Fourth Annu Meeting Soc Magn Reson Med 806
97. Wyrwicz AM, Murphy R, Prakash I, Moriarty RM, Dougherty I (1985) Fourth Annu Meeting Soc Magn Reson Med 4: 827
98. Berkowitz BA, Ackerman JJH (1985) Fourth Annu Meeting Soc Magn Reson Med 759
99. Babcock EE, Nunnally RL (1985) Fourth Annu Meeting Soc Magn Reson Med 751
100. Fox JL (1984) Science 224: 143
101. Keniry M, Benz C, Shafer RH, James TL (1986) Cancer Res 46: 1754

# Measuring Cation Movements Across the Cell Wall Using NMR Spectroscopy: Sodium Movements in Striated Muscle

**Joanne S. Ingwall**

NMR Laboratory for Physiological Chemistry, Department of Medicine, Brigham and Women's Hospital, Harvard Medical School, 221 Longwood Avenue, Boston, Massachusetts 02115, USA

## Table of Contents



New applications of NMR spectroscopy of intact tissues allow repetitive, sequential measurement of the intracellular amount or concentration of several cations important for cell function: $Na^+$, $K^+$, $H^+$, $Ca^{2+}$ and $Mg^{2+}$. Intra- and extracellular amounts of $Na^+$ or $K^+$ can be measured simultaneously using $^{23}Na$ or $^{39}K$ NMR by placing a frequency shift reagent in the extracellular

NMR Basic Principles and Progress, Vol. 28

space; a comparison of three anionic chelate complexes of paramagnetic lanthanides used as shift reagents is presented. Intracellular $Ca^{2+}$ concentrations may be measured using $^{19}F$ NMR and the $Ca^{2+}$ chelator 5F-BAPTA. Intracellular pH is esimated from the chemical shift dependance of the inorganic phosphate resonance in the $^{13}P$ NMR spectrum. By placing phenylphosphonic acid in the extracellular space, extracellular pH may also be estimated from its chemical shift. Also using $^{31}P$ spectra, concentration of free $Mg^{2+}$ may be inferred from the relative chemical shifts of the [α-P] and [β-P] ATP resonances and the dissociation constant of MgATP. From these measurements, movement of cations across the cell wall may be assessed. Representative examples using striated muscle are presented.

Emphasis is made on the use of frequency shift reagents to discriminate between $^{23}Na$ signals arising from the intra- and extracellular spaces, including analysis of the source of the shifted and unshifted signals, field dependence of the shift and contributions from bulk magnetic susceptibility shift effects. Information derived from $^{23}Na$ spectra obtained for the gastrocnemius muscle of the living rat, including $Na^+$ movements during ischemia, is described to illustrate application of $^{23}Na$ NMR with shift reagent to biological samples.

# 1 Introduction

## *1.1 Importance of Cation Gradients for Cell Function*

Intracellular cations play important roles in cell division and maintaining normal cell function in both prokaryotic and eukaryotic cells. Plasma membranes function to maintain large concentration gradients of ions, notably the monovalent cations sodium ($Na^+$), potassium ($K^+$) and proton ($H^+$) and the divalent cations calcium ($Ca^{2+}$) and magnesium ($Mg^{2+}$), thereby regulating thier intracellular concentrations. A good example of the role of cation movements in normal cell function is the control of excitation-contraction coupling in striated muscle: During a twitch in skeletal muscle and during each cardiac cycle, movement of $Na^+$, $Ca^{2+}$, $H^+$ and $K^+$ (as well as other charged species) across the plasma membrane (referred to as the sarcolemma in muscle cells) combine to initiate a sequence of events leading to contraction: $Na^+$ and $Ca^{2+}$ move from the interstitium to the intracellular space while $K^+$ and $H^+$ move out of the cell. Muscle relaxation occurs when the concentrations of these critical cations (most importantly $Ca^{2+}$) return to baseline. The source of energy for these cation movements is the hydrolysis of ATP, either directly via enzymes which stoichiometrically couple ATP hydrolysis and ion translocation or indirectly through the sodium gradient. Cytosolic concentrations of $Na^+$, $K^+$, $H^+$ and $Ca^{2+}$ are determined by the integrated action of ion pumps, exchangers and channels of all cellular membranes: the sarcolemma, the sarcoplasmic reticulum and the inner mitochondrial membrane.

*Cation Movement Across the Sarcolemma*: In the cytoplasm of normal mammalian striated muscle, $Na^+$, $Ca^{2+}$ and $H^+$ exist in concentrations below electrochemical equilibrium (Table 1, with definitions): thus energy consuming processes are required to extrude them from the cell. $[K^+]_i$ is higher than electrochemical equilibrium; it leaks from the cell and ATP is required to pump it back to the cytoplasm. The processes that maintain normal chemical gradients of these cations across the sarcolemmal membrane form a complex and interdependent network of pumps, channels and exchangers (Fig. 1). A summary of the primary components of this network follows.

**Table 1.** Extra- and intracellular cation concentrations $[X]_0$, $[X]_i$ (in mM), their ratio and equilibrium concentrations for striated muscle

| | $[X]_0$ | $[X]_i$ | $[X]_0/[X]_i$ | $[X]_i$ equil[a] |
|---|---|---|---|---|
| $Na^+$ | 145 | 5 | 29 | 4100 |
| $K^+$ | 5 | 145 | 0.03 | 141 |
| $Ca^{2+}$ | 1 | $10^{-4}$ | $10^4$ | 0.8 |
| $H^+$ | $4 \times 10^{-5}$ | $7 \times 10^{-5}$ | 0.56 | $10^{-3}$ |

[a] $[X]_i$ calculated for equilibrium assuming an electrical potential of $-90$ mV at 37 °C

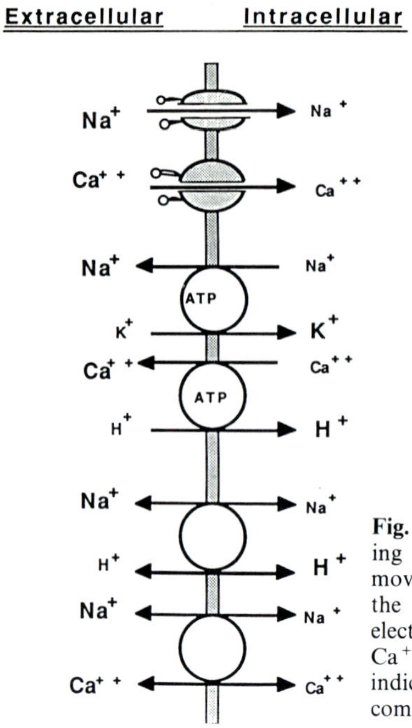


**Fig. 1.** Diagramatic representation of the sarcolemma showing the primary pathways of $Na^+$, $Ca^{2+}$, $K^+$ and $H^+$ movement: the voltage dependent $Na^+$ and $Ca^{2+}$ channels, the ATP-dependent sodium and calcium pumps, the electroneutral $Na^+$, $H^+$ exchange and the electrogenic $Na^+$, $Ca^+$ exchanger. The relative size of the symbol for each ion indicates the relative difference in concentations in the two compartments

**$Na^+$**: Efflux of $Na^+$ from the myocyte depends on $Na^+$, $K^+$-ATPase activity (the sodium pump), which ejects 3 $Na^+$ ions in exchange for 2 $K^+$ ions with the hydrolysis of 1 ATP molecule. Major pathways of $Na^+$ influx are via the fast sodium (voltage-dependent) channel; influx may also occur via the reversible $Na^+$, $H^+$ and $Na^+$, $Ca^{2+}$ exchangers. The direction of exchange depends on the gradients for each ion, stoichiometry of the exchange and the membrane potential [1, 2]. Other pathways such as $Na^+$, $K^+$, $Cl^-$ cotransport and Na-dependent phosphate, glucose and alanine cotransporters may be active under some conditions.

**$Ca^{2+}$**: Intracellular calcium ion, $Ca_i^{2+}$, is extruded by the high-affinity, low capacity, ATP-dependent $Ca^{2+}$, $2H^+$ pump. When $[Ca^{2+}]_i$ is high, as during a twitch in skeletal muscle or during systole and early diastole in heart, $Ca^{2+}$ efflux occurs primarily by the low affinity, high capacity $Na^+$, $Ca^{2+}$ exchanger. The sarcolemmal $Na^+$, $Ca^{2+}$ exchanger is electrogenic (3 $Na^+$ in, 1 $Ca^{2+}$ out) and thus is driven by both chemical and electrical transmembrane gradients [1–4]. $Ca^{2+}$ influx occurs via the voltage-dependent $Ca^{2+}$ channels. It is subject to control by cAMP-dependent phosphorylation and is inhibited by high $[H^+]_0$ (5). Depending on the membrane potential, $Ca^{2+}$ influx also occurs via reserved $Na^+$, $Ca^{2+}$ exchange.

**$H^+$**: Efflux of $H^+$ occurs via the electroneutral $Na^+$, $H^+$ exchanger (among other processes including phosphate and lactate efflux). The amiloride-sensitive $Na^+$, $H^+$ exchanger may play a major role in regulating intracellular pH and thus cell volume [6–8]. One pathway for $H^+$ influx is via the $Ca^{2+}$ pump (1 $Ca^2$ out, 2 $H^+$ in).

**$K^+$**: $K^+$ Efflux may be passive or ATP-dependent [9]. $K^+$ influx occurs primarily via the Na pump.

*Cation Movement Across Organellar Membranes*: The sarcoplasmic reticulum in muscle cells contains $Ca^{2+}$ channels allowing $Ca^{2+}$ efflux to the cytosol and a high-affinity $Ca^{2+}$, $H^+$ ATPase which provides the energy for rapid $Ca^{2+}$ uptake. Similarly; mitochondria accumulate $Ca^{2+}$ via a low affinity, high capacity uniporter; $Ca^{2+}$ efflux occurs via an electroneutral $Na^+$, $Ca^{2+}$ exchanger. $K^+$ and $H^+$ movements across the mitochondrial membranes also occur [10].

The integration of these pathways for cation influx and efflux occurs in all tissues, although their relative adundance and physical properties differ [11]. Because of their importance in our understanding of normal and pathophysiological function of tissues, measuring their cytosolic concentrations and their movements via specific pathways in healthy and diseased states has been the subject of many investigations.

## *1.2 Classical Methods for Measuring Cation Content*

Non-NMR tools used to measure cytosolic cation content and their movements across membranes include flame photometry, atomic absorption, electron probe X-ray microanalysis, ion-selective electrodes, ion-specific fluorescent dyes and chemilumiscent proteins and use of radioisotopes.

Many measurements of cation content for whole tissue have been made using flame photometry and atomic absorption. Since most cations are present in high concentrations in either the extra- or intracellular spaces, these measures report primarily values for one, but not both, of these compartments. Exhaustive flushing of the extracellular space improves on the accuracy of these measurements, but concerns about incomplete removal of ions from the extracellular space and resultant perturbation of intracellular ion contents have not been overcome. Electron probe X-ray microanalysis provides elegant measurement of ion density with resolution at the electron micrographic level. Importantly, each of these techniques measures the amount of ion present, not its concentration. Separate measurement, using other techniques, of the water contents of the intra- and extracellular spaces is required to calculate concentrations. The volumes of the intra- and extracellular compartments are highly dependent on experimental conditions, and are known with variable degrees of accuracy.

The development of ion-selective microelectrodes has permitted direct measurement of the chemical activities of several cations in cells, notably $Ca^{2+}$, $Na^+$ and $H^+$. Electrodes for intracellular $K^+$ are less reliable. Activity is directly related to concentration by means of the thermodynamic parameter, the activity coefficient; this coefficient depends on the ion's environment and hence can differ between the intra- and extracellular spaces. Calibration of these electrodes is often problematic, but, because of their sensitivity, ion- selective electrodes have provided important data defining changes in intracellular $Na^+$, $H^+$ and $Ca^{2+}$ and extracellular $K^+$ activities in many cell types [for examples 12, 13]. Importantly, unlike the destructive methods described above, measurements using microelectrodes can be made continuously, reporting changes that occur in response to perturbation. However, these techniques can be applied only to quiescent muscles. Successful application to muscle has been confined to myocytes, muscle strips and the surface cells of whole muscle, but not the intact functioning muscle.

In common with the advantages and limitations of ion-selective electrodes, the bioluminescent indicator aequorin [14] and the fluorescent indo-1 and fura-2 compounds [for example, 15] are sensitive tools for the continuous measurement of intracellular $Ca^{2+}$ but report ion concentration only in cells, thin muscles or in small regions of intact tissue. Other fluorescent compounds have been used to measure intracellular concentrations of $H^+$ with good success, and for $Na^+$ but with more limited success. Ideally, these reporter molecules should (1) have fast response time, (2) load in low enough concentrations so as not to buffer the cation pool being studied, (3) not be sequestered in intracellular organelles and (4) be readily standardized. The aequorin technique satisfies points 1–3 but is difficult to calibrate. Fluorescent compounds for $Ca_i^{2+}$ have slow kinetics, buffer, sequester in intracellular organelles, and are difficult to calibrate. Thus, there is no ideal indicator for measurement of ion concentrations in vivo. Importantly none of these techniques provide simultaneous measurement of intra- and extracellular cation pools.

## *1.3 NMR Methods for Measuring Cation Content and Movements In Vivo, with Representative Applications Using Striated Muscle*

$H^+$, $Na^+$, $K^+$, $Mg^{2+}$ and $Ca^{2+}$ each have nuclides which undergo nuclear magnetic resonance (Table 2 and Ref. 16) The properties of high natural abundance, relatively high concentrations in biological tissues and short values

**Table 2.** NMR properties of cationic nuclides

| Nuclide | Spin quantum number | Natural abundance (%) | NMR[a] receptivity | $T_1$ (msec) | Frequency at 8.4 T (MHz) |
|---|---|---|---|---|---|
| $^{1}H$ | 1/2 | 100 | 100 | $10^2$–$10^3$ | 360 |
| $^{23}Na$ | 3/2 | 100 | 9.27 | ~$10^0$ | 95.2 |
| $^{39}K$ | 3/2 | 93.1 | 0.048 | ~$10^0$ | 16.8 |
| $^{25}Mg$ | −5/2 | 10.13 | 0.027 | ~$10^2$ | 22.0 |
| $^{43}Ca$ | −7/2 | 0.145 | 0.00087 | ~$10^3$ | 24.2 |

[a] From Ref. [12]

for spin-lattice relaxation times combine to make $Na^+$ and $K^+$ (but not $Mg^{2+}$, $Ca^{2+}$ or $H^+$) amenable to direct study in biological samples using NMR. Estimates of $H^+$, $Mg^{2+}$ and $Ca^{2+}$ contents by NMR must be made using indirect techniques.

As will be reviewed below, the unique characteristic of NMR measurement of $Na^+$, $K^+$, $Ca^{2+}$, $H^+$ and $Mg^{2+}$ contents is that repetitive non-destructive measurements can be made for intact functioning tissue. Moreover, many of the techniques allow simultaneous measurement of extra- and intracellular ion content, thereby permitting quantitative assessment of *net* ion movement. With few exceptions, NMR reports the amount of ion, not the concentration, requiring separate measurements of cell water to calculate the thermodynamically relevant entity concentration.

### *1.3.1 $Na^+$ and $K^+$*

Due to its high natural abundance, high concentration in biological samples, short $T_1$ values and isochronicity, $^{23}Na$ is the second (next to $^1H$) most NMR-sensitive nucleus found in biological tissues. The first $^{23}Na$ NMR spectrum of biological sample was obtained in 1956 for blood [17]; the first $^{23}Na$ image of a biological sample was obtained in 1981 for perfused heart [18]. The chemical shifts of the $^{23}Na$ NMR signals for $Na^+$ in the intra- and extracellular spaces are essentially indistinguishable. To discriminate between these signals, two approaches have been used. The first attempts to use characteristics inherent in the NMR experiment, using no perturbations of the biological system other than imposed by $B_0$ and $B_1$ fields. Since nuclei with spin 3/2 are described by four nuclear spin states, defining three single-quantum NMR transitions (Fig. 2 from Ref. 19), more than one value for relaxation constant may exist. Some investigators [20] have explored the possibility that multiple quantum NMR could detect differences in relaxation rates of intra vs extracellular nuclei. Although there is evidence that more than one rate constant is required to describe observed relaxation [21], it is also clear that cations present in both the intra and extracellular spaces contribute to both relaxation constants. Thus, quantitative analysis of $Na^+$ or $K^+$ signals arising from Na or K distributed in either intra or extracellular spaces cannot be made using this approach.

The second approach is to use frequency shift reagents. In the early 1980s, three groups independently showed that the apparent degeneracy of the three $^{23}Na$ NMR transitions could be perturbed by using membrane-impermeable shift reagents [22–24]. Dysprosium tripolyphosphate, $Dy(PPP)_2^{7-}$ was developed by Gupta and Gupta [23]; triethylenetetraminehexaacetatedysprosium (III), $DyTTHA^{3-}$ by Springer and colleagues [25] and 1,4,7,10-tetra-azacyclododecane-*N*,*N*′,*N*″,*N*‴-tetra(methylenephosphonate) thulium · (III), $TmDOTP^{5-}$, by Sherry, Malloy and their colleagues [26]. Their use successfully discriminates between $^{23}Na$ and $^{39}K$ NMR signals arising from the intra- and extracellular spaces in a wide variety of biological samples ranging from single

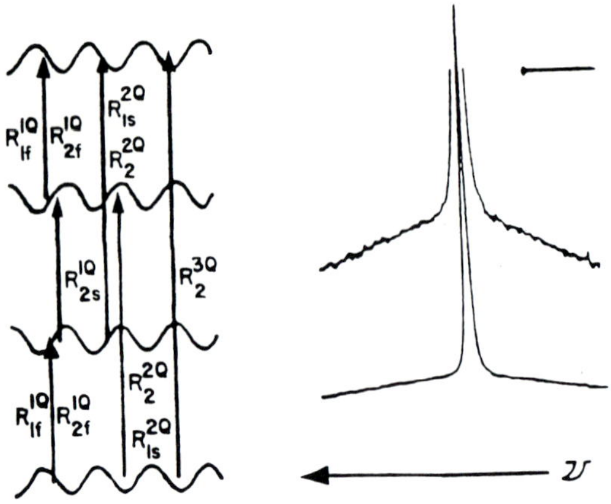

**Fig. 2.** Rotating frame energy level diagram for isolated spin 3/2 system in fast exchange. The type c spectrum is for $Na^+$ in an aqueous solution which contains a high concentration of micelle-solubilized gramicidin channels. It is originally Fig. 5 of Urry DW, Tapane TV, Venkatachalam CM, McMichens BM (1989) Meth Enzymol 171: 342 and was used by Rooney and Springer [19] to assign the most probable scheme for $Na^+$ in a biological environment: homogeneous biexponential with a narrow central transition (1/2 to $-1/2$ transition) superimposed on two satellite coherences (3/2 to 1/2 and $-1/2$ to $-3/2$)

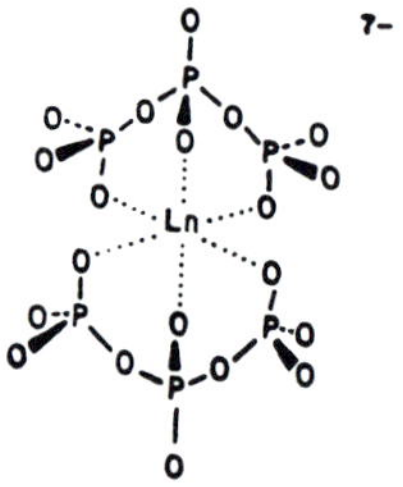

$Dy(PPP)^{7-}$
dysprosium tripolyphosphate

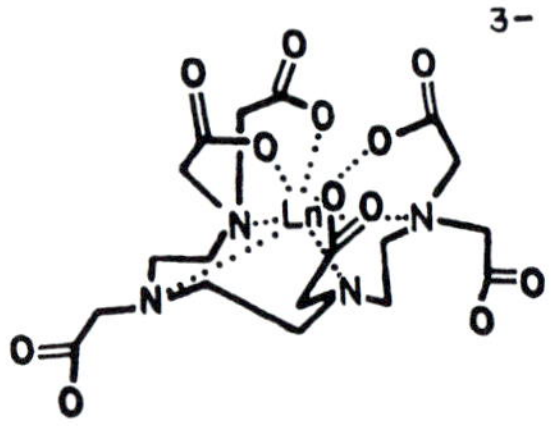

$DyTTHA^{3-}$
triethylenetetraminehexaacetatedysprosium

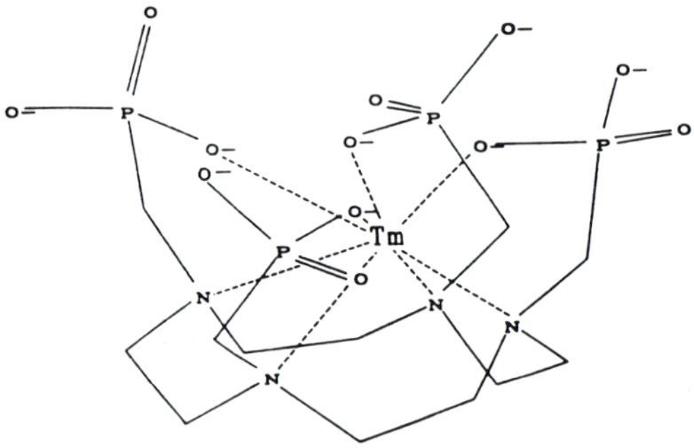

$TmDOTP^{5-}$
1,4,7,10 - tetraazacyclododecane - N,N',N'',N''' - tetra (methylenephosphonate) thulium

**Fig. 3.** Putative structures for the three shift reagents currently in use to discriminate intra- and extracellular $Na^+$ and $K^+$ in biological systems. From references [25] and [52]

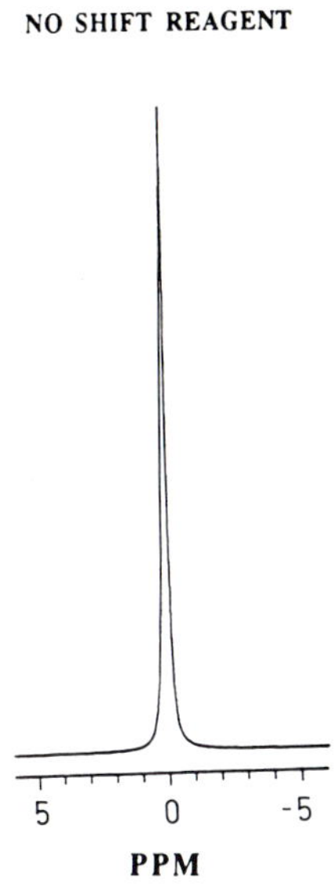


**Fig. 4.** $^{23}$Na spectra of isolated isovolumic rat heart (~ 1.25 g wet weight) perfused either with no shift reagent (**top**) 10 mM $DyTTHA^{3-}$ (**left**) or 10 mM $TmDOTP^{5-}$ (**right**) in phosphate-free Krebs-Henseleit buffer containing 118 mM NaCl, 4.7 mM KCl, 1.75 mM $CaCl_2$, 1.2 mM $MgCl_2$, 0.5 mM $Na_2H_2EDTA$, 25 mM $NaHCO_3$ and 11 mM D-glucose. For spectra shown for hearts perfused with shift reagent, hearts were simultaneously bathed in isotonic containing shift reagent. Perfusate flow was ~ 20 ml/min and mannitol flow was 1.5 times greater. Spectra were obtained at 8.45 T using a Nicolet NT-360 wide-bore NMR spectrometer (8.45) operating at 95.25 MHz for $^{23}$Na. Shimming was performed on the $^{23}$Na signal using an 18 Channel Oxford Instruments Shim Supply. $^{23}$Na NMR spectra were obtained using a pulse angle of 90° and a recycle time of 250 ms. Spectra consisted of 480 FIDs, providing time resolution of 2 min. Spectral width was ±2 kHz and 1 K data points were collected. The position of $^{23}$Na signal without shift reagent was set to 0 ppm

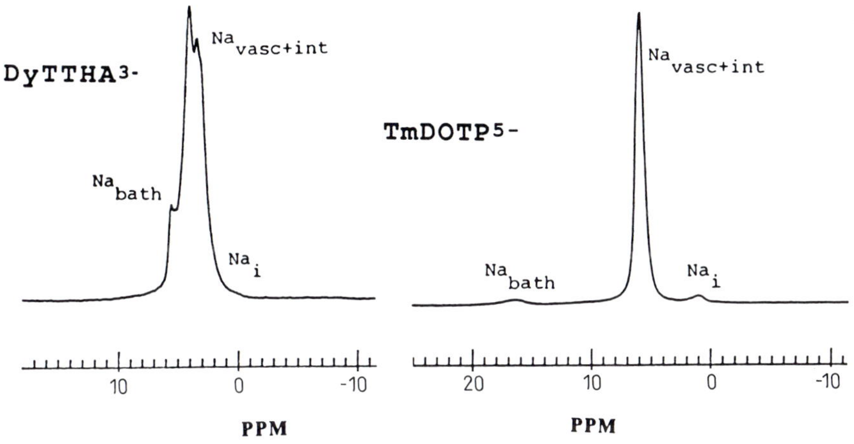


cells such as yeast [27] and red blood cells [28] to isolated perfused tissues such as heart [24] to organs within the intact animal [29, 30]. The putative structures of these three shift reagents are shown in Fig. 3. Figure 4 shows spectra obtained for an isolated rat heart perfused either without shift reagent (top spectrum), with the shift reagent $DyTTHA^3$ (middle spectrum) or with the shift reagent $TmDOTP^{5-}$ (bottom spectrum).

Analogous considerations apply to measurement of $K_i$ and $K_o$. However, low frequency and hence low sensitivity have slowed its application. The first $^{39}$K NMR signal from tissue was made in 1970 [29]; the first in muscle in 1985 [24]. Figure 5 shows a two $^{39}$K spectra for isolated perfused rat heart, one

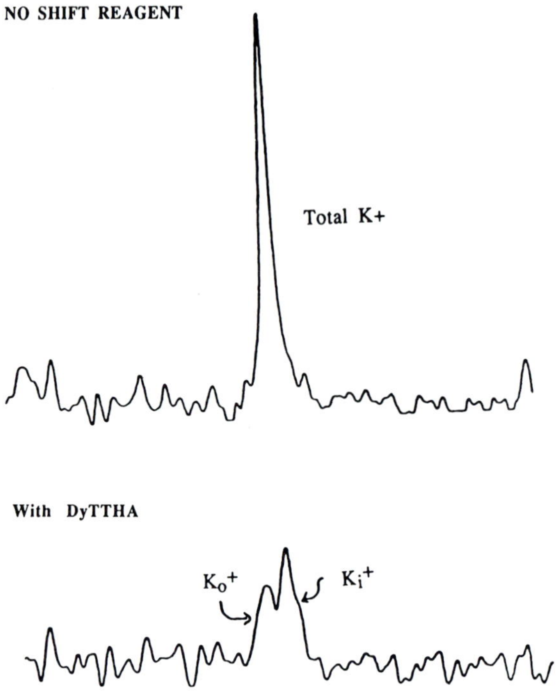


**Fig. 5.** $^{39}K$ spectra is isolated isovolumic rat heart; details as in Fig. 3. The shift reagent was 10 mM $DyTTHA^{3-}$. A Nicolet wide-bore NT-360 MHz (8.45 T) NMR spectrometer (Nicolet Magnetics Corp., Spectra were acquired with 90° pulses (140 μs) and an interpulse delay of 256 ms (940 accumulations; 4 min time resolution). The sweep width was ±4000 and 2 K data points were collected. Spectra were obtained by Dr. L. C. Stewart

showing isochronous signal and one showing how the presence of shift reagent $DyTTHA^{3-}$ in the extracellular spaces allow discrimination of signals corresponding to intrta- and extracellular $K^+$ pools. Use of frequency shift reagents is the focus of Sects. 2 and 3.

### *1.3.2* $Ca^{2+}$

Direct measurement of intracellular $Ca^{2+}$ in biological samples by $^{43}Ca$ NMR is not practical due in part to the prohibitive cost of the NMR sensitive nuclide. However, several investigators have used NMR techniques to measure $[Ca^{2+}]_i$ indirectly. In 1983, Metcalfe and colleagues [32] introduced a new NMR method for the quantitation of intracellular $Ca^{2+}$ in biological systems. The method uses fluorinated 1,2-bis(*o*-aminophenoxy)ethane-*N*,*N*,*N'*,*N'*-tetraacetic acid (nF-BAPTA) (Fig. 6), which complexes with $Ca^{2+}$ in both slow and fast exchange, to report $[Ca^{2+}]_i$ via $^{19}F$ NMR. The compound is delivered as the acetoxymethyester (5F-BAPTA-AM) (5 μM). Intracellular hydrolysis produces

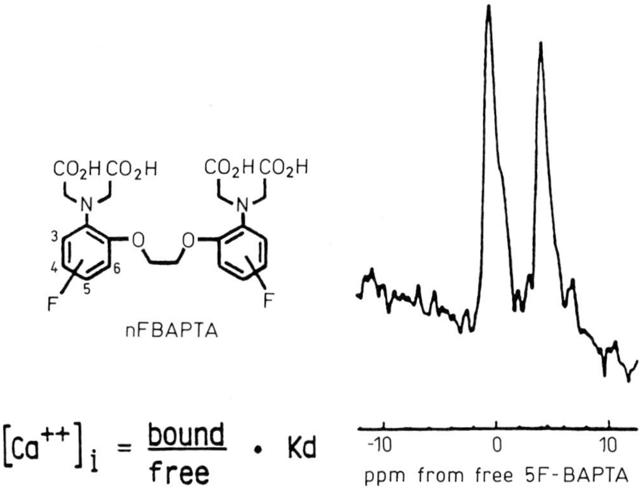


**Fig. 6.** Structure of nF-BAPTA and $^{19}$F NMR spectrum obtained for perfused rat heart taken from Ref. 32. $[Ca^{2+}]_i$ was estimated to be 690 nM using a $K_D$ of 709. See text and reference for details

an impermeable acid which accumulates to ~0.1 to 0.6 mM. Extracellular 5F-BAPTA-AM and 5F-BAPTA are washed away leaving only intracellular 5F-BAPTA. $Ca^{2+}$-bound and free 5F-BAPTA have distinct $^{19}$F NMR resonances (Fig. 6); the ratio of resonance areas for bound to free 5F-BAPTA times the dissociation constant reports $[Ca^{2+}]_i$. Several groups have now reported results for $[Ca^{2+}]_i$, in heart using this method [32–35], although the value of Kd is still in dispute.

The method is attractive for several reasons. $^{19}$F is almost as sensitive as $^1$H and without natural background. A recent report by Marban and colleagues [34] carefully documents its characteristics and distribution with the cell. The method is not without it problems, however. These include the large volumes needed to load the muscle and existence of multiple $Ca^{2+}$ pools in the myocyte. A major problem with this technique is that the intracellular concentration of 5F-BAPTA is so high that the $Ca_i^{2+}$ pool is perturbed. A side product of acetoxymethy de-esterification is the formation of formaldehyde. The uniform white color of rat hearts and patchy white regions observed in ferret hearts perfused with 5F-BAPTA may reflect cell necrosis resulting from formaldehyde toxicity. Whether due only to $Ca_i^{2+}$ buffering or to a combination of toxicity and $Ca_i^{2+}$ buffering, the normal amplitude of contraction in 5F-BAPTA-loaded is greatly diminished. The possibility that the fundamental characteristics of the pressure trace may be changed has not been addressed. Attempts to overcome inhibition of contractile function by increasing $[Ca^{2+}]_o$ to 8 mM have been only partly successful [34, 35]. On the other hand, the observation by Steenbergen et al. [33] and Marban et al. [34] that the free cytosolic $[Ca^{2+}]_i$ in resting heart measured using this technique is in good agreement with other methods is encouraging. In an important application of this technique to heart, Steenbergen et al. [33] report that $Ca^{2+}$ measured this way increases ~5-fold between 9 and 15 mins after ischemia, and that this accumulation is reversible. In another, Marban and colleagues [35] used this technique to define the magnitude of the $Ca^{2+}$ transient during the cardiac cycle: $[Ca^{2+}]$ increased from 300 nM during diastole to ~1000 nM during systole.

### *1.3.3 $H^+$*

$^{31}$P NMR spectroscopy may provide the only truly non-invasive method for measuring intracellular pH. The first report using $^{31}$P NMR to measure $pH_i$ was made in red blood cells by Moon and Richards [36]. The sensitivity of the chemical shift of inorganic phosphate to intracellular pH is based on the phosphate equilibrium: $H_3O^+ + HPO_4^{2-} \rightleftharpoons H_2PO_4^- + H_2O$ which has a pKa near 6.85. Simple expressions relating pH to the differences in chemical shift of these species have been formulated [for a review, see 37]. The relationship between chemical shift and pH can be determined empirically using a standard curve, taking care to mimic the ionic strength of the intracellular milieu. Operationally, changes in splitting between the phosphocreatine and inorganic phosphate resonances are measured.

Since most cells, tissues and organs contain a NMR-detectible level of inorganic phosphate (>0.5 mM in 1 g sample at 8.4 T), intracellular pH can usually be estimated. It is widely accepted that the value of $pH_i$ in perfused tissue is close to 7.1 and falls to pH 6–6.2 in tissue injured by cessation of flow (ischemia). In vivo, the resonances for blood-borne 2,3-diphosphoglycerate (DPG) may occlude the intracellular Pi signal. Selectively saturating the downfield resonance of 2, 3-DPG effectively diminishes intensity of both resonances, allowing measurement of $pH_i$ [38]. A large extracellular Pi signal can also be problematic and is the reason why most cell and isolated organ experiments are performed using Pi-free buffers. This has no untoward effects on high energy phosphate content during normal perfusion.

Direct estimation of extracellular pH may also be measured repetitively using $^{31}$P NMR by placing the disodium salt of phenylphosphonate (PPA) which has a pKa near 7.07, in the extracellular space. This compound is effectively excluded from the intracellular space. By adding the compound to the buffer perfusate of isolated bladder [39], skeletal muscle [40] and heart [41], simultaneous assessment of $pH_o$ and $pH_i$ has been made. In heart and skeletal muscle preparations, the difference between $pH_i$ and $pH_o$ is 0.24 pH unit. Figure 7 shows a $^{31}$P NMR spectrum of an isolated rat heart perfused with PPA. Clarke et al. [41] studying isolated rat heart have shown that this difference is abolished within minutes after imposing perfusate flow ceased. Thus, the ratio of $[H^+]_i/[H^+]_o$ changed from 1.7 to ~1.

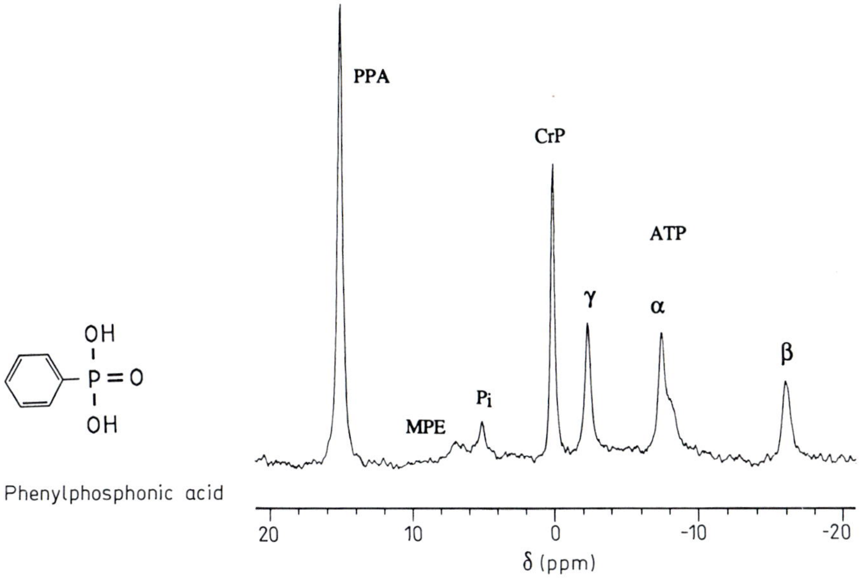


**Fig. 7.** 31P NMR spectrum of isolated isovolumic rat heart perfused with phosphate-free Krebs-Henseleit buffer (see legend to Fig. 4) containing 15 mM phenylphosphonic acid. Spectra were obtained using a Nicolet NT-360 wide bore NMR spectrometer operating at 145.75 MHz for $^{31}P$. The spectrum was obtained by signal averaging 104 FIDs using 60° broad-based pulse with a 2.15 s recycle time (total time 4 min), spectral width was $\pm 3$ kHz and 2 K data points were collected

The signal of PPA is proportional to the volume distribution of PPA; thus, $^{31}P$ NMR spectroscopy with PPA also provides information about changes in extracellular volume of a tissue. By simultaneously supplying PPA and dimethyl methylphosphonate, which distributes into the total water space, total and extracellular volumes may be measured and, by difference, the intracellular volume can be estimated. Such on-line measurements provide essential information needed to convert NMR-measured amounts into concentrations. This technique has been applied to the isolated rat heart [42].

### *1.3.4 $Mg^{2+}$*

The position of the [$\beta$-P] ATP resonance of the $^{31}P$ NMR spectrum depends on the ratio of MgATP and all ATP species (proton and Mg-bound, as well as free) present in the sample. $[Mg^{2+}]_i$ can be calculated from the difference between the chemical shifts of [$\alpha$-P] and [$\beta$-P] ATP resonance using the relationship:

$$[Mg^{2+}]_i = K_D MgATP(\phi^{-1} - 1) \qquad (1a)$$

where

$$\phi = \frac{[ATP]_{free}}{[ATP]_{total}} = \frac{\delta_{\alpha\beta} - \delta_{\alpha\beta MgATP}}{\delta_{\alpha\beta ATP} - \delta_{\alpha\beta MgATP}} \quad (1b)$$

where $\delta_{\alpha\beta}$ is the difference in chemical shift between the [$\alpha$-P] and [$\beta$-P] ATP resonances in the sample, and $\delta_{\alpha\beta ATP}$ and $\delta_{\alpha\beta MgATP}$ are the separations for ATP in solution with $Mg^{2+}$ absent or present at saturating concentrations.

The method is generally attributed to Gupta and colleagues who applied the technique to red blood cells [43] and skeletal muscle [44]. In skeletal muscle ~93% of ATP is MgATP and free $Mg^{2+}$ is 0.2–0.6 mM [44]. More recently, several groups have applied the technique to measurement of free $Mg^{2+}$ in heart, reporting a value of ~0.8 mM in control heart and 2–3-fold increases in the presence of inotropic agents [45] or ischemia [46].

## 2 Identifying Signals in $^{23}$Na NMR Spectra Obtained for Biological Samples Supplied with Shift Reagent

The basic requirements for a frequency shift reagent suitable for use in biological samples are (1) that it induces a shift in nuclear spin resonance frequency sufficient to allow discrimination of signals arising from intra- and extracellular cation pools and (2) that it does not perturb normal physiology of the system. Thus, one must consider both NMR and biological properties while designing shift reagents.

### *2.1 Hyperfine Shift and Bulk Magnetic Susceptibility Shifts*

Shift reagents developed for measurement of cation content in biological samples are anionic chelate complexes of paramagnetic lanthanides, specifically $Dy^{3+}$ and $Tm^{3+}$ (Fig. 3). The presence of the paramagnetic lanthanide causes the hyperfine shift of the resonance frequency of nuclides transiently bound to the shift reagent. It occurs because of the physical interaction between magnetic dipoles of cation and the lanthanide's unpaired electronic density while the cation is transiently bound to the shift reagent. The presence of the lanthanide also produces bulk magnetic susceptibility (BMS) shifts, perturbing $^{23}$Na NMR signals arising from $Na^+$ in the same and nearby compartments. The anionic character of shift reagents serves two important functions: (1) it allows interaction with cations and (2) it minimizes binding to biological macromolecules and to membranes, essentially precluding its entry into the cell.

The chemical shift of a nucleus is determined by contributions from host molecule electronic shielding effects, solvent electronic shielding effects, and BMS effects. In general, perturbation of the chemical shift induced by the presence of a paramagnetic compound in the same compartment includes a diamagnetic contribution due to complex formation, contact hyperfine shift and dipolar hyperfine (pseudocontact) shift; solvent effects cancel. When the shift reagent is not uniformly distributed (as in biological samples), BMS effects also contribute. In the case of shift reagents, the cation does not bond directly to the lanthanide; thus, the contact hyperfine shift does not contribute. The complex formation shift term is also negligible. Thus, the chemical shift induced by the presence of shift reagent is due primarily to dipolar hyperfine shift and BMS shift. Each will be considered.

The shift in the $^{23}$Na resonance frequency caused by the presence of the shift reagent (illustrated here by $DyTTHA^{3-}$) is, to a first appoximation, a consequence of the following simple equilibrium:[13]

$$Na^{+} + DyTTHA^{3-} \leftrightharpoons NaDyTTHA^{2-}.$$

Since this equilibrium is fast on the NMR time scale, the resonance frequency $\Delta$ of the observed $^{23}$Na signal is the weighted average of the resonance frequency of free $Na^{+}$ (defined as zero) and that of the $Na^{+}DyTTHA^{2-}$ complex ($\Delta^{\circ}$, 159 ppm at 25°C). Thus, $\Delta = f_b \Delta^{\circ}$, where $f_b$ is the fraction of $Na^{+}$ which is bound to $DyTTHA^{3-}$. For the equilibrium above,

$$f_b = \frac{[NaDyTTHA^{2-}]}{[Na^{+}] + [NaDyTTHA^{2-}]}. \qquad (2)$$

Since $\Delta^{\circ}$ for $DyTTHA^{2-}$ is downfield, the observed shift becomes more downfield as $f_b$ becomes larger. The value of $f_b$ depends, in turn, upon the concentrations of $Na^{+}$ and $DyTTHA^{3-}$, temperature (as does $\Delta^{\circ}$), the ionic strength of the medium, and the concentrations of cations which complete with $Na^{+}$ for $DtTTHA^{3-}$ (particularly $Ca^{2+}$ and, to a lesser extent, $Mg^{2+}$) and anionic sites that compete with $DyTTHA^{3-}$ for $Na^{+}$. In the aqueous milieu in vivo, it is reasonable to assume that all of these quantities except $[DyTTHA^{3-}]$ are similar in every compartment accessible to shift reagent, i.e., the interstitial and vascular spaces.

Orientation of the sample relative to $B_0$ determines the lineshapes of both the shifted and unshifted resonances. An important and illustrative example of this was presented by Springer and colleagues [47] using parallel coaxial cylinders. The center tube was filled with $Dy^{3+}$, $DyTTHA^{3-}$ or $Dy(PPP)_2^{7-}$ and NaCl; the outer one with only NaCl. $^{23}$Na spectra obtained when the long axis of the phantom was placed either perpendicular (top panel) or parallel (bottom panel) to $B_0$ are shown in Fig. 8. The profound contributions of local susceptibility effects are shown by the spectra obtained when $Dy^{3+}$ was present in the center tube. In this case, no hyperfine shift can occur; the shifts observed are due only to BMS effects. The unpredictability of the direction and magnitude of BMS effects is also illustrated: 3.8 ppm splitting downfield for parallel orientation in contrast to 1.9 ppm splitting for perpendicular orientation. Comparison of the spectra obtained when shift reagent was present show that, for either shift reagent, orientation relative to $B_0$ changes the extent of shift,

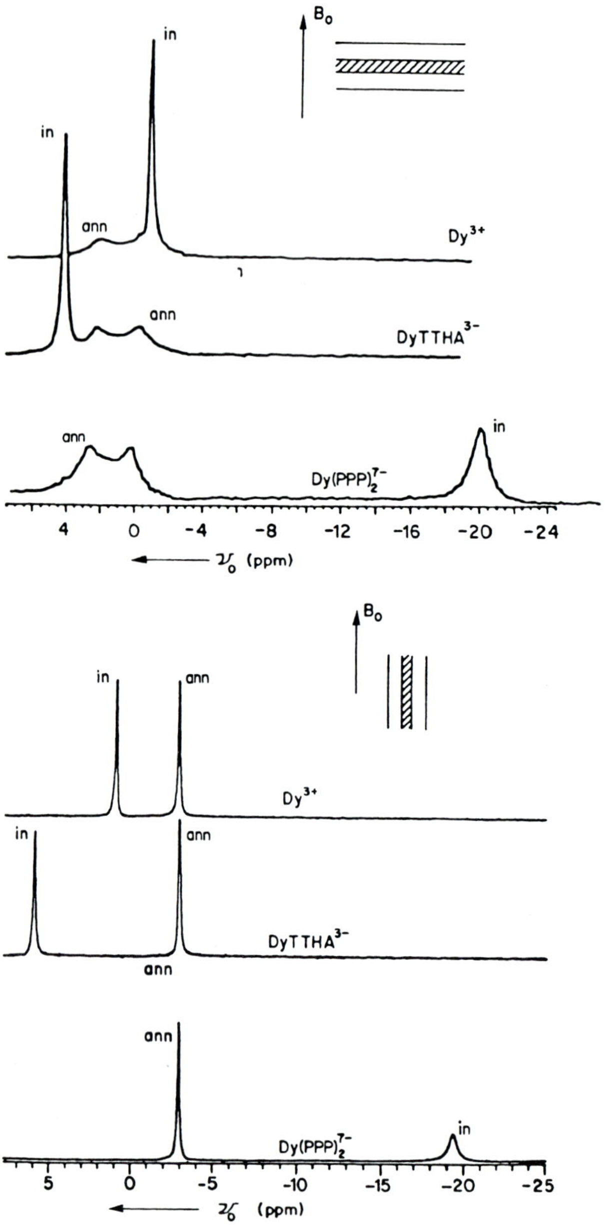


**Fig. 8.** $^{23}Na$ spectra obtained for coaxial tubes containing NaCl and either $Dy^{3+}$, $DyTTHA^{3-}$ or $Dy(PPP)_2^{7-}$ in the inner tube (represented by the *hatched area*) and only NaCl in the outer tube. Spectra on the top were obtained when the coaxial tubes were placed perpendicular to the magnetic field ($B_0$); spectra on the bottom when the tubes were placed parallel to $B_0$ (79.4 MHz). Redrawn from [47]

line width and lineshape: sharp resonances were obtained for parallel configuration, but broad, irregularly shaped resonances were obtained for Na even in the shift reagent-free annulus when the long axis of the phantom is perpendicular to $B_0$. Thus, $^{23}$Na spectra of a multicompartment system in which only some of the compartments contain shift reagent are determined by combination of hyperfine shift and BMS shifts. The BMS shifts are not predictable but depend on differences in the susceptibility, shape and orientation of each compartment with respect to the magnetic field. Based on this analysis using non-biological phantoms, we may now consider biological samples.

The simplest biological sample is a suspension of cells. Based on its uniformity in terms of shape and orientation to $B_0$, the $^{23}$Na spectrum of yeast cells suspended in shift reagent-containing buffer should reflect primarily hyperfine shift with little BMS shift effects. Figure 9 taken from Rooney and Springer [19] compares the direction and extent of the hyperfine shifts caused by suspending Na-loaded yeast in buffer containing equal concentrations (9 mM) of either $Dy(PPP)_2^{7-}$ or $DyTTHA^{3-}$. $DyTTHA^{3-}$ shifts the $Na_o$ signal to a higher frequency while $Dy(PPP)_2^{7-}$ shifts the $Na_o$ signal to a lower frequency; $Na_i$ signal is relatively unshifted in either case. Importantly, the extent of shift is four times larger for $Dy(PPP)_2^{7-}$; the linewidth of the shifted signal is also about four times larger. As expected, the perturbation in lineshape due to BMS effects is small.

A biological sample closely approximating the co-axial phantom of Springer and colleagues [47] is skeletal muscle. It contains parallel fibers containing

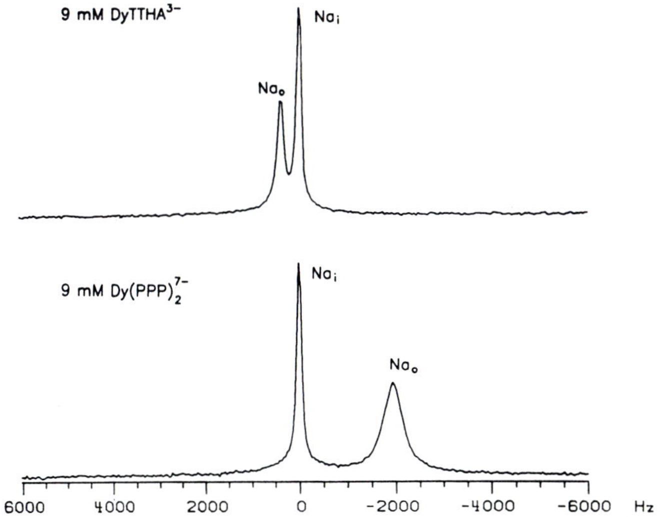


**Fig. 9.** $^{23}$Na NMR spectra of Na-loaded yeast suspended in buffer (45% cell density) containing either 9 mM $DyTTHA^{3-}$ or 9 mM $Dy(PPP)_2^{7-}$ at 37 °C. The peak at 480 Hz in the top panel corresponds to extracellular Na; in the lower panel, extracellular Na resonance is at −2190 Hz. From Ref. [19]

proteins which are also arranged in a fixed geometry parallel to the long axis of the fiber. Thus, changing the orientation of a piece of excised skeletal muscle superfused with shift reagent in the magnetic field should reveal any BMS contributions to the $^{23}$Na spectrum. Figure 10, taken from Kohler et al. [48] shows spectra for the same piece of gastrocnemius muscle bathed in $DyTTHA^{3-}$ for four different orientations relative to the magnetic field. The relatively unshifted $Na_i$ resonance is not discernible on this scale, but clearly seen is the dependence of the shape and linewidth of the shifted $Na_0$ signal on orientation. Without analyzing perturbations due to BMS shifts in this way, one might erroneously assign what appears to be three signals to three physical compartments (in the bottom spectrum), instead of one compartment whose signal is affected by BMS shifts.

Another example, taken from the same report [48], emphasizes the need to develop a strategy to identify the physical source of the signals, i.e. to distinguish between BMS effects and the presence of multiple compartments. In this experiment (Fig. 11), an isolated rat heart was submerged in isotonic mannitol-containing $DyTTHA^{3-}$ while it was perfused with buffer containing shift reagent; thus, only the intracellular compartment did not contain $DyTTTA^{3-}$. The $Na_i$ signal is large because the heart was subjected to no flow for 28 min. The bottom $^{23}$Na spectrum shows a large relatively unshifted resonance near zero ppm and a complex set of resonances between 2–5 ppm. The heart was

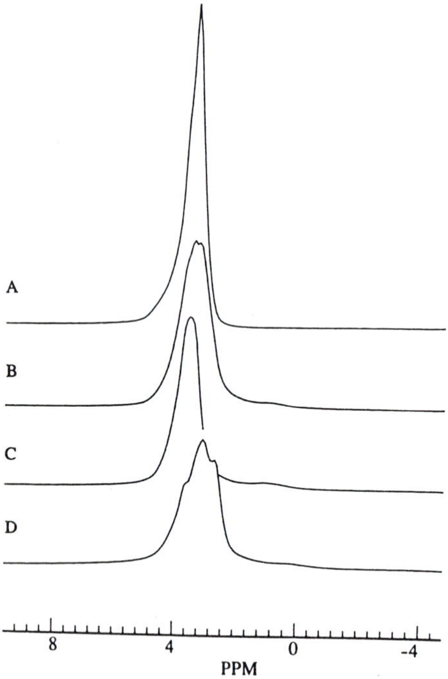


**Fig. 10.** $^{23}$Na spectra (95 MHz) of a sample of gastrocnemius muscle bathed in shift reagent-containing buffer. Spectra A–D were taken as the location of the muscle in the sample tube was varied and muscle/solution ratio was varied. A small intracellular sodium signal may be seen at approximately 0.5 ppm. From Kohler et al. [48]

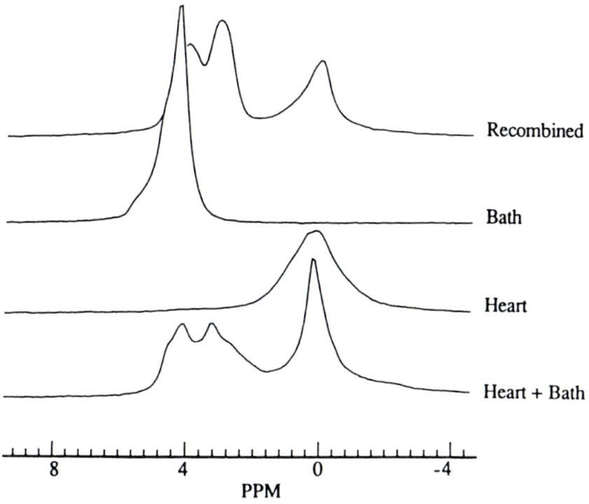


**Fig. 11.** $^{23}Na$ spectra of a rat heart perfused with shift reagent-free buffer and bathed by a mannitol solution containing shift reagent $DyTTHA^{3-}$. Spectra are shown of the heart and bath together, the heart alone, the bath alone, and the recombined system. Physiological and NMR details are as described for Fig. 4. From Kohler et al. [48]

then removed and spectra were obtained for the heart alone, the bathing fluid alone and then for the recombined heart plus bath. This simple experiment clearly identifies the physical sources of the signals and illustrates the contributions of BMS shift effects on the shifted signals. A parallel experiment in which $^{1}H$ spectra were obtained for heart alone, bath alone and then recombined showed that no signal could be assigned to intracellular water and that the inhomogenity in the $^{1}H$ signal is due exclusively to BMS shift effects. This was expected since the ratio of $DyTTHA^{3-}$ to water is several orders of magnitude too small to produce a detectable shift. However, several investigators have erroneously assigned this small upfield $^{1}H$ 'peak' to an intracellular compartment [49, 50].

## 2.2 *Field Dependence of Shift*

$^{23}Na$ signals reporting interstitial and intracellular $Na^{+}$are inhomogenously broadened in the $^{23}Na$ spectra at all field strengths. Balschi et al. [51] obtained $^{23}Na$ spectra at 1.5, 4.7 and 8.4 T for gastrocnemius muscle of the living rat infused with $DyTTHA^{3-}$. When plotted on an ppm scale, linewidth showed little field dependence, with or without shift reagent. This experiment illustrates two important points. The first is that the observed linewidths are determined primarily by BMS shifts. The second is that the shift reagent technique can be successfully used to discriminate $Na_o$ and $Na_i$ at the low fields commonly used for human MRI and MRS.

## 2.3 *Choosing a $^{23}$Na NMR Shift Reagent for Biological Applications: Balancing NMR and Biological Constraints*

Each of the shift reagents currently in use to distinguish $Na_0$ and $Na_i$ in biological samples produces spectra with both hyperfine and BMS shifts, but their relative contributions differ (Fig. 4). The example for yeast cells in Fig. 9 comparing $Dy(PPP)_2^{7-}$ and $DyTTHA^{3-}$ show that, for the same concentration of shift reagent, the hyperfine shift is much larger for $Dy(PPP)_2^{7-}$ than for $DyTTHA^{3-}$. Conversely, as shown in Fig. 4 for isolated rat heart, the contribution of BMS shift effect is smaller for $TmDOTP^{5-}$ than for $DyTTHA^{3-}$, making areas of $Na_i$ and $Na_o$ resonances easier to quantitate. For the same concentration, the shift is almost twice as large with $TmDOTP^{5-}$ than for $DyTTHA^{3-}$. Thus, based on NMR characteristics, $Dy(PPP)_2^{7-}$ would be ranked first in terms of shift, followed by $TmDOTP^{5-}$ and then by $DyTTHA^{3-}$. On the basis of broadening effects, $TmDOTP^{5-}$ is superior. Importantly, at concentrations of 3–10 mM each of these shift reagents shifts the $Na_o$ signal (145 mM) at least a 2–4 ppm, allowing discrimination of $Na_o$ and $Na_i$ signals.

Shift reagents must also be compatible with physiological requirements of the system under study. While each shift reagent shown in Fig. 3 should be useful for experiments using cell suspensions, each is not equally compatible with intact physiological systems. Of particular importance is the compatibility of shift reagent with physiological levels of free $Ca^{2+}$ in blood or in perfusates designed to mimic ion content of blood. The minimum free $[Ca^{2+}]_o$ needed to maintain normal excitation-coupling in excitable tissues is $\sim$1.2 mM. $Dy(PPP)_2^{7-}$ precipitates with $Ca^{2+}$, yielding solutions in which maximum free $[Ca^{2+}]$ is $\sim$0.5 to 0.7 mM. $TmDOTP^{5-}$ binds $Ca^{2+}$ with close to 1:1 stoichiometry; free $Ca^{2+}$ is limited to $\sim$1 mM for solutions containing equal concentrations of $Ca^{2+}$ and $TmDOTP^{5-}$. 10 mM $DyTTHA^{3-}$ on the other hand, requires only 3 mM ionized $Ca^{2+}$ to yield a free $Ca^{2+}$ of 1.2 mM. Thus, $DyTTHA^{3-}$ and probably $TmDOTP^{5-}$ are acceptable, while $Dy(PPP)_2^{7-}$ is not.

Another requirement is that, for most experiments, the shift reagent must not perturb normal physiologic and biochemical functions. The compound must be stable; any free lanthanide would be toxic. By assessing cardiac performance (heart rate, aortic pressure, left ventricular pressures) and the contents of the high energy phosphate compounds, ATP and phosphocreatine, short-term toxicity can be assessed. Subject to action by phosphatases, $Dy(PPP)_2^{7-}$ is not stable in most biological samples. As a result free lanthanide and pyrophosphate may exert toxic effects; $DyTTHA^{3-}$ and $TmDOTP^{5-}$ are stable in physiologic buffers and blood. Neither $DyTTHA^{3-}$ nor $TmDOTP^{5-}$ [52] disturb heart function or high energy phosphate content in isolated hearts. Supplying $DyTTHA^{3-}$ (maximum concentration of 8.8 mM in blood) to the living rat had no effects on the hemodynamics or high-energy phosphate content of skeletal muscle [30]; a comparable in vivo test of $TmDOTP^{5-}$ has been reported stating that it has acute toxicity [53]. Importantly both $DyTTHA^{3-}$ and $TmDOTP^{5-}$

allow essentially simultaneous acquisition of $^{23}Na$ and $^{31}P$ NMR spectra. Interleaved spectra have been obtained for the gastrocnemius muscle of the living rat [30] and for isolated hearts first using $DyTTHA^{3-}$ [24] and more recently using $TmDOTP^{5-}$ [54]. In skeletal muscle perfused with $DyTTHA^{3-}$, $^{31}P$ resonance areas obtained in interleaved fashion were unaffected; their positions were shifted by only ~0.25 ppm, due to BMS effects. Preservation of cardiac function and high energy phosphate content also support the conclusion that the sarcolemma is not permeable to these shift reagents, thereby allowing discrimination of $Na^+$ in the intra- vs extracellular spaces.

Identification of the source of the Na signals in the $^{23}Na$ spectrum is based on the results using non-biological phantoms, simple biological systems and reconstitution experiments such as described above. In addition, assignments have been made by observing expected changes in putative $Na_o$ and $Na_i$ signals. For example, Pike et al. [24] showed that the putative $Na_i$ signal increased when the sodium pump in hearts was inhibited with ouabain or with perfusion with 0 mM $K^+$. Neubauer et al. [55] observed increases in $Na_o$ during hypoxia when vessels are known to expand and contain more perfusate Na. Partitioning the interstitial and vascular Na signals that comprise the "extracellular" Na pool requires physical maneuvers such as collapse of the vessels during zero-perfusion pressure or replacing vascular Na with $Li^+$ or choline [48]. Taken together, there is no doubt that shift reagent can discriminate between intra- and extracellular $Na^+$ pools without perturbation of the intracellular milieu.

## 3 Application of $^{23}Na$ NMR Spectroscopy with Shift Reagent to Skeletal Muscle in the Living Rat: Defining Sodium Movements During Ischemia

Each Na signal has three important properties: Chemical shift, intensity and lineshape. Lineshape has been discussed in Sect. 2. Here information obtained from chemical shift and intensity are discussed. The system chosen to illustrate the points is resting skeletal muscle of the living rat made ischemic by total occlusion of the femoral vein artery and nerve bundle [30]. It is chosen for three reasons. First, vascular volume in resting muscle is small; thus, the muscle can be modeled as a two compartment system with $Na^+$ exchange occurring between interstitium and intracellular compartments. Second, literature values for the volumes of these spaces are in good agreement, making calculations of $Na_o$ and $Na_i$ concentration possible using the NMR-measured amount. Third, total ischemia produces a system which is essentially closed, trapping shift reagent in the muscle, where interstitial $Na^+$ must move down its concentration gradient into the intracellular space. The experimental protocol has been

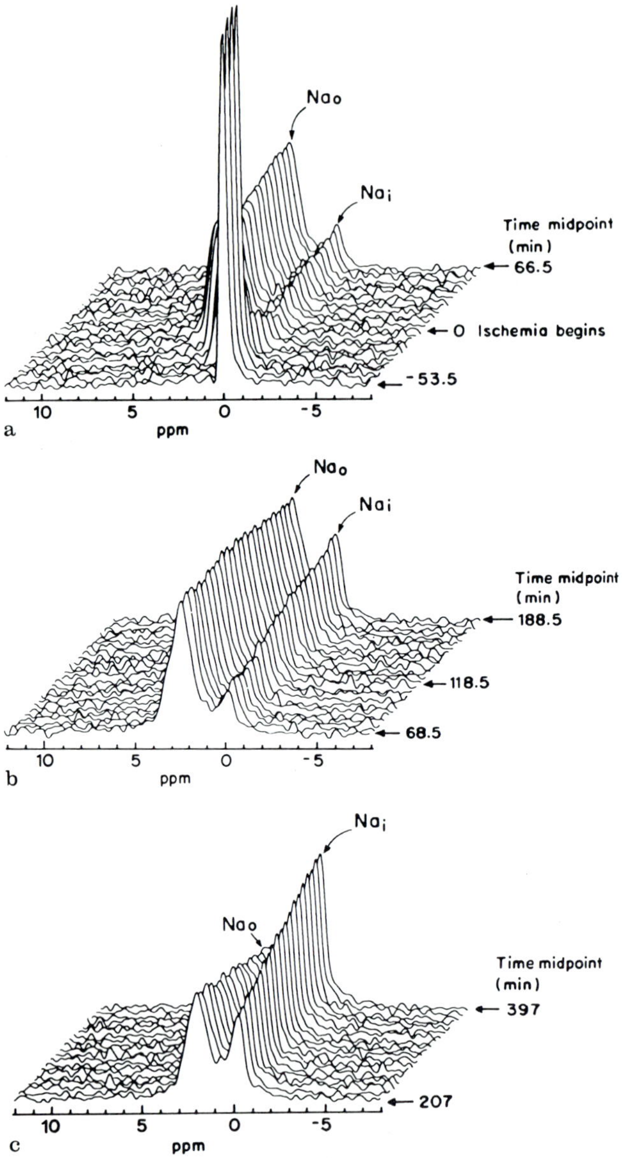


**Fig. 12 a–c.** $^{23}$Na NMR spectra of normal and ischemic gastrocnemius muscle of the living rat obtained using a 2-turn 11 mm surface coil at 8.45 T obtained before and during ischemia. **Panel a** shows spectra obtained during shift reagent infusion (total of 1 mmol $DyTTHA^{3-}$ infused in ~ 275 g rat) and the pre-ischemic period (t = − 53.5 to 0) and during the first hour of ischemia and after shift reagent infusion had ceased (t = 0 to 66.5 min). Note that the $Na_0$ peak was shifted ~ 26 ppm; $Na_i$ shifted upfield ~ 0.25 ppm due to BMS effects. **Panel b** shows $^{23}$Na NMR spectra obtained during the second and third hour of ischemia. **Panel c** shows $^{23}$Na NMR spectra obtained after 3 hours of ischemia. The time between the start of each spectrum is 10 min. Each resolution enhanced spectrum is the sum of 600 scans obtained over 2 min; times indicated are the midpoints of the acquisitions. From Balschi et al. [30]

described [30]. Representative spectra showing changes in $^{23}Na$ signal of the gastrocnemius muscle during $DyTTHA^{3-}$ infusion and during subsequent period of no flow are shown in Fig.12.

## 3.1 *Information Derived from Changes in Chemical Shift*

Information derived from changes in chemical shift of $Na_o$ resonance includes: (1) kinetics of shift reagent permeation into the muscle and its washout, (2) extent of perfusion and (3) distribution volume of shift reagent.

The concentration of $DyTTHA^{3-}$ in the interstitium of resting skeletal muscle in the living rat depends on blood flow and the instantaneous concentration of shift reagent in the blood perfusing the muscle. The concentration of shift reagent in blood depends upon the rate of shift reagent infusion into the jugular vein, the rate of permeation of $DyTTHA^{3-}$ into the interstitial spaces of all tissues of the body, and the rate of clearance of the shift reagent from the bloodstream via the urinary tract. Thus, for a constant rate of infusion, the interstitial spaces of well-perfused organs of the body, such as the heart and kidney, will equilibrate with the shift reagent in the blood much more rapidly than the interstitial space of resting gastrocnemius muscle. Since the concentration of blood $Na_o$ during shift reagent infusion into a living rat changes only slightly relative to changes in shift reagent concentration, the chemical shift of $Na_o$ signal is determined primarily by shift reagent concentration. Balschi el al. [30] have shown that loss of shift reagent from muscle (due to excretion) can be modeled as a pseudo-first order process. By plotting shift vs time (Fig.13), pseudo-first order rate constants describing loss of shift reagent during normal

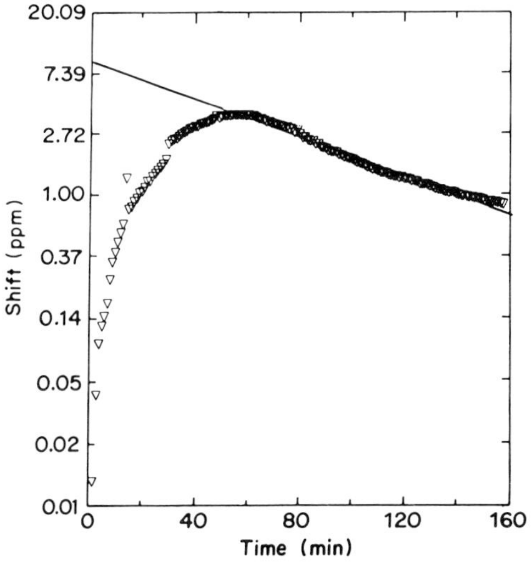


**Fig. 13.** Chemical shift of the $Na^+$ signals obtained from the well-perfused gastocenemius of the rat loaded with 8 mM $DyTTHA^{3-}$ and then allowed to clear through the urinary tract. By modelling the loss of shift reagent as a first order process, the rate constant for clearance can be calculated. The semilogarithimic plot of the shift of $Na_0$ vs time fields the rate constant; the straight line of equation, in $y = \ln 9.02 - 0.0158\,t$, $(r2 = 0.98)$, is fit only to those points measured after the end of shift reagent infusion ($t = 57$ min). During total ischemia, the rate constant is $\sim 10^{-3}$ $min^{-1}$. Values in between define partial ischemia. From Balschi et al. [30]

perfusion and total ischemia were calculated. For normal well-perfused muscle, this rate constant was $0.0158\,\mathrm{min}^{-1}$ and the half-life for loss of shift reagent from the muscle was 43 min. These values compare favorably to time of wash-in (assessed by constancy in both the $Na_i$ signal intensity and in the chemical shifts of $Na_i$ and $Na_o$) of 40 min and half-life of compound with similar structures Gd DPTA of 59 min. They are also in close agreement with measured rates of excretion [56]. As expected, during total ischemia, the chemical shift of $Na_o$ changed only slightly (0.7 ppm in 250 min), yielding a pseudo-first order rate constant of $\sim 1 \times 10^{-3}\,\mathrm{min}^{-1}$ and a half-life of excretion of $> 16$ hour. Importantly, this analysis shows that this approach can be used to assess the extent of remaining blood flow during partial ischemia.

The pseudo-first order model also provides information about the volume distribution of shift reagent. The intercept of the shift vs time relationship (Fig. 13) corresponds to the extrapolated value of the shift of $Na_o$ as if all the shift reagent that was infused were still in the body (bolus injection model). Using the relationship between chemical shift and shift reagent concentration obtained empirically for blood, the concentration of shift reagent was estimated. By knowing the amount infused, the volume was directly calculated. The distribution volume of 15% (% of total body weight) calculated by using $^{23}Na$ NMR in this way compares favorably with classical measurements of 16.5% (12% interstitial plus 4.5% plasma) [57]. This close agreement provides compelling evidence that this shift reagent does not enter the intracellular space. This technique should have application to other experimental settings.

## *3.2 Information Derived from Changes in $^{23}Na$ Signals*

The $Na_i$ signal is proportional to the amount of unbound $Na_i$ in that space(s) not accessible to shift reagent. Based on the evidence reviewed above, we conclude that this includes the entire intracellular volume and no other Na-containing compartment. Electron probe X-ray microanalysis of heart muscle confirms [58] the assumption that almost all cytosolic $Na_i^+$ is unbound and also provides evidence that $Na^+$ is homogeneously distributed through the intracellular space: At least for chick heart myocytes, the concentration of Na in mitochondria is the same as in the cytoplasm. The nuclear envelope and sarcoplasmic reticulum are probably permeable to $Na^+$ as well. Thus, Na is most likely homogeneously distributed in the intracellular space (this does not preclude the possibility of gradients existing near membranes).

The amount of $Na_{o,i}$ measured by $^{23}Na$ NMR is the volume-weighted average over the entire cell population contained within the NMR-sensitive volume. For many tissues, there is significant heterogeneity of cell type; however, for striated muscle, myoctes comprise 80% or more of tissue volume. Thus, changes in putative $Na_i$ signal faithfully report changes in the amount of $Na_i$ in myocytes.

### 3.2.1 *"Visibility Factors"*

The amount of NMR-measured Na in any compartment is proportional to the product of its concentration, the volume of the compartment and its so-called "NMR-visibility" factor. Thus, if volume and visibility factor were known, $^{23}$Na signals could be used to calculate cation concentrations and hence Na movements across the cell wall. For this reason, the history of $^{23}$Na NMR is replete with attempts to establish both the NMR visibility factor of $Na_i$ and to offer explanations for NMR "invisibility". Although equally important, assessment of volume has not received as much attention.

Published visibility factors for $Na^+$ in biological systems range from 0.1 to 1.0. It is widely held that for a system in which all the Na is the aquo ion and the ion is in a simple bulk solution (such as the vascular spaces), the three single quantum transitions (Fig. 2) are degenerate and the visibility is 1.0. In contrast, for at least some biological systems, it has been thought that only the central transition is "visible", yielding a visibility factor of 0.4. Other have suggested that each of the transitions were equally visible or invisible and thus the signal from each transition must contribute to the observed signal. In an elegant theoretical and experimental analysis of all six relaxation rate constants of a $Na^+$ resonance of a single population of isolated spins in fast exchange, Rooney and Springer [19] have recently shown that a major determinant of the so-called "visibility" is the delay between the hard pulse and enabling of the receiver, the"dead time". The shorter the dead time, the higher the visibility factor, i.e. the more of the signal that will be captured. This dependence may explain some of the variability in literature values. Both theory and experiment show that in the limiting case for delays of $\sim 300\,\mu$sec, 40% of the signal may be undetected. They also identify the most likely spectral form for Na in a biological environment where Na encounters primarily negatively changed macromolecules during its diffusional excursions. They provide evidence that biological Na spectrum is type c, or homogeneous biexponential: narrow central resonance (1/2 to 1/2 transition) superimposed on two satellite coherences (3/2 to 1/2 and $-1/2$ to $-3/2$) (Fig. 2).

The skeletal muscle experiments being used as the model system analyzed here show that the visibility factors of the interstitial and intracellular Na are essentially the same, or more precisely, that the visibility of the $Na^+$ which moves from the intestitium to the intracellular spaces is unchanged. This is based on data shown in Fig. 14 plotting the areas of $Na_i$ and $Na_o$ and their sum vs time for leg muscle supplied with shift reagent by infusion into the intact rat followed by total ischemia (for spectra shown in Fig. 12). The initial increase in total Na during shift reagent infusion is due to supply of exogenous $Na^+$ as counterion of the shift reagent. The observation that the total Na signal during ischemia does not change very much (within 15% when corrected for a small amount of edema) even though there large changes in $Na_o$ and $Na_i$ signal intensities means that the visibility of the moving Na ion does not change.

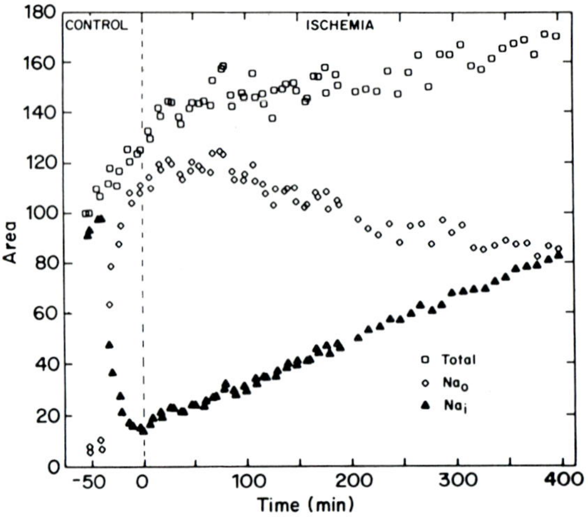


**Fig. 14.** Area of the Na signals obtained from ischemic muscle. Plots of the $^{23}$Na NMR integrated areas of $Na_0$, $Na_i$, and $Na_t$ vs time for the experiment shown in Fig. 12. The total area of the first spectrum (t = − 53.5 min in Fig. 12) was arbitrarily set to 100 units. From Balschi et al. [30]

An analogous analysis of the $Na^+$ signals for the vascular compartment of an isolated perfused heart and its surrounding bath during collapse of the vasculature (caused simply by changing perfusion tissue from 100 to 0 mmHg) showed that the visibility of $Na^+$ moving from vascular to the bath is not altered by changing environments [59]. Thus, it is likely that the visibility factors are similar for interstitial and intracellular spaces (probably 1.0) and for vascular and bath (probably less than 1.0). By using standard curves, carefully constructed to mimic lossey environments, coupled with classical techniques, Bernard et al. [60] found that the visibility factors for intracellular and interstitial spaces in heart is ~ 0.66, a value close to that predicted from the analysis of Rooney and Springer [19].

### *3.2.2 Calculation of $[Na^+]_{i,o}$*

There are two ways to calculate $Na_i$ concentrations using $^{23}$Na spectra, the ratio method and external standard method. The ratio method takes advantage of the simultaneous measurement of $Na_o$ and $Na_i$ signals in $^{23}$Na spectra with shift reagent.

Assuming that the NMR-visibility of the $^{23}$Na signal is the same in the extra- and intracellular compartments, the intracellular $Na^+$ concentration can be calculated using the relationship:

$$\frac{A_{Na_i}}{A_{Na_o}} = \frac{[Na^+]_i V_i}{[Na^+]_o V_o} \tag{3a}$$

where $A_{Na_o}$ = shifted peak area, $A_{Na_i}$ = unshifted resonance area, $[Na^+]_i$ = intracellular $Na^+$ concentration, $[Na^+]_o$ = extracellular (interstitial) $Na^+$ concentration, $V_i$ = intracellular volume, and $V_o$ = extracellular volume. Solving for $[Na^+]_i$ yields:

$$[Na^+]_i = \frac{A_{Na_i}}{A_{Na_o}} \frac{V_o}{V_i} [Na^+]_o. \tag{3b}$$

For a 3-compartment system where the compartments do not have equal visibilities, as is the case for a tissue with a substantial vascular bed, the equation is more complicated:

$$[Na^+]_i = \frac{A_{Na_i}}{A_{Na_o}} \frac{(V_v \upsilon_v + V_{is} \upsilon_{is})}{(V_i \upsilon_i)} [Na^+]_o \tag{4}$$

where $V_{is}$ = interstitial volume, $V_v$ = vascular volume, $v_i$ = intracellular $Na^+$ signal NMR visibility, $\upsilon_{is}$ = interstitial $Na^+$ signal NMR visibility and $\upsilon_v$ = vascular $Na^+$ signal visibility.

The ratio method is appropriate for surface coil experiments where the NMR-sensitive volume has an irregular shape. In experiments where the sample is entirely contained in the NMR-sensitive volume, an external or reference standard may be used to relate area to amount. In this case, the visibility of the compartments reported by $Na_o$ and $Na_i$ must be known.

In both cases, the volume of each compartment must be known. In the case of resting skeletal muscle, there is good literature agreement. However, for isolated buffer-perfused organs, there is little or no information about the magnitude of the vascular bed during perfusion or how it changes with time and intervention. The promising work of Clarke and colleagues [41, 42] using the external $^{31}P$-observable marker PPA (see Sect. 1) to calculate total and extracellular water spaces should solve this problem.

Results from both NMR and classical experiments show that $[Na^+]_i$ differs for different striated muscles for a given species: 5 mM for rat skeletal muscle by NMR [30] and 10 mM by neutron activation analysis [61] vs 15 mM for rat heart by both ion-selective electrodes [13] and NMR [60]. For the same tissue, $[Na^+]_i$ also differs among species: for heart, $[Na^+]_i$ in hearts of rat, rabbit and guinea pig are 15 mM [13, 60], 10 mM [13] and ~10 mM [60] respectively. A more complete comparison of $[Na^+]_i$ determined by classical techniques and by $^{23}Na$ for heart has been recently summarized [62]. The $Na_o$ signal is proportional to the sum of sodium in the vascular space and interstitium. Although differences in Na concentration in these two extracellular compartments may exist, they must be small.

### *3.2.3 $Na^+$ Movements Across the Cell Wall*

Changes in $Na_i$ and $Na_o$ signal areas during ischemia in the gastrocnemius muscle of the living rat shown in Figs. 12 and 14 can be used to assess movement of $Na^+$ from extra- to the intracellular spaces. Since the ischemic leg is a closed two-compartment system, cation movement can be modeled as a pseudo-first order process. The rate constant(s) for $Na^+$ entry may be calculated by fitting the time dependence of the ln of the ratio of $Na_o$ at any time divided by the initial $Na_o$. If it is truly a two-compartment system, the same rate constants will describe the increase in $Na_i$ relative to starting conditions. Inspection of the relations of signal areas to time in Fig. 14 shows that two rate constants are required to describe $Na^+$ movement during ischemia, one for the first 10 min of ischemia, and another for later times. The challenge now is to delineate how the components of the interconnected network shown in Fig. 1 change during early vs late ischemia (or any other perturbation under study). One approach is to define the basic thermodynamic descriptions of each pathway. By using NMR-measured values for sarcolemmal cation gradients of $Na^+, K^+, H^+$, and $Ca^{2+}$, coupled with estimates of the membrane potential approximated by the potassium equilibrium potential values for change in free energy for coupled ion movements can be calculated: For the $Na^+, K^+$ATPase reaction and for $Na^+$, $H^+$ and $Na^+$, $Ca^{2+}$ exchanges. For example, let us consider the electroneutral $Na^+$, $H^+$ antiporter for the reaction $Na_o^+ + H_i^+ \leftrightharpoons Na_i^+ + H_o^+$,

$$\Delta G_{Na,H} = RT \ln \frac{[Na^+]_i\,[H^+]_i}{[Na^+]_o\,[H^+]_o}. \tag{5}$$

Using NMR-measured values for $[Na^+]_{i,o}$ and $[H^+]_{i,o}$ during normal perfusion and ischemia to calculate the sign and magnitude of $\Delta G_{Na,H}$, we learn that $Na^+$, $H^+$ exchange always functions to exchange intracellular $H^+$ for extracellular $Na^+$, even in ischemia. Using NMR-measured $H^+$ and $Na^+$, several groups [11, 62] have used pharmacologic perturbation of $Na^+$, $H^+$ exchange in intact heart to assess relative contribution of $Na^+, H^+$ exchange to $Na_i^+$ accumulation in ischemia. Such approaches as these should have application to other tissues and organs.

*Acknowledgments*: I would like to acknowledge my many collaborators who, over the past seven years, have worked to develop the field of in vivo cation NMR spectroscopy: Paul Allen, David Atkinson, Marianna Bak, Monique Bernard, James A. Balschi, John A. Bittl, Kieran Clarke, April Dutta, Jonathan Frazer, Susan Kohler, Stefan Neubauer, Stanton B. Perry, Martin M. Pike, Thomas W. Smith, Charles S. Springer, Jr. and Laura C. Stewart. In particular, I acknowledge the pioneering theoretical and experimental contributions made, and still being made, to this field by Charles S. Springer, Jr. and his influence on our work.

## 4 References

1. Eisner DA, Luderer WJ (1985) Am J Physiol 248: C189
2. Mullins LJ (1985) Ion transport in the heart, 000, 000
3. Carafoli E, Penniston JT (1985) Sci Amer 253: 70

4. Reeves JP (1985) Current Topics in Membranes and Transport 25: 77
5. Langer GA (1985) Circ Res 57: 374–382
6. Lazdunski M, Frelin C, Vigne P (1985) J Mol Cell Cardio 17: 1029
7. Seifter JL, Aronson PS (1986) J Clin Invest 78: 859
8. Piwnica-Worms D, Jacob R, Horres Cr, Leiberman M (1985) J Gen Physiol 85: 43
9. Weiss JN, Lamp ST (1987) Science 238: 67
10. Brierly GP (1983) Pathobiology of Cell Membranes III: 23
11. Bernard M, Ingwall JS (1990) Circulation 82: III–686
12. Kleber AG (1983) Cir Res 52: 442
13. Shattock MJ, Bers MD (1989) Amer J Physiol 256: C813
14. Pesaturo JA, Gwathmey JK (1990) Circ Res 66: 696
15. Leé HC, Smith N, Mohabir R, Clusin WT (1987) Proc Natl Aca Sci (USA) 252: C441
16. Harris RK (1983) In: Nuclear magnetic resonance spectroscopy: A physicochemical view. Pittman, Marshfield MA: p 230
17. Jardetsky O, Wertz (1956) Amer J Physiol 187: 608
18. DeLayre JD, Ingwall JS, Malloy C, Fossel ET (1981) Science 212: 935
19. Rooney WD, Springer CS (1991) Magn Reson Med (in press)
20. Lyon RC, Pekar J, Moonen CTW, McLaughlin AC (1991) Magn Reson Med (in press)
21. Burstein D, Litt HI, Fossel ET (1989) Magn Reson Med 9: 66
22. Degani H, Bar-On Z (1981) Period Biol 83: 61
23. Gupta RK, Gupta P (1982) J Magn Res 47: 344
24. Pike MM, Frazer JC, Dedrick DF, Ingwall JS, Allen PD, Springer CS Jr, Smith TW (1985) Biophys J 48: 159
25. Chu S C-K, Pike MM, Fossel Et, Smith TW, Balschi JS, Springer CS Jr (1984) J Magn Reson 56: 33
26. Sherry AD, Malloy CR, Jeffrey FMH, Cacheris WP, Geraldes FGC (1988) J Magn Reson 76: 528
27. Hofeler H, Jensen D, Pike MM, Delayre JL, Cirillo V, Springer CS Jr, Fossel ET, Balschi JA (1987) Biochemistry 26: 4953
28. Pike MM, Fossel ET, Smith TW, Springer CS Jr (1984) Amer J Physiol 246: C528
29. Blum H, Schnall MD, Chance B, Buzby GP (1988) Amer J Physiol 255: C377
30. Balschi JA, Bittl JA, Springer CS Jr, Ingwall JS (1990) NMR Biomed 3: 47
31. Cope FW, Damadian R (1970) Nature 228: 76
32. Metcalfe JC, Hesketh RT, Smith GA (1985) Cell Calcium 6: 1983
33. Steenbergen C, Murphy E, Levy L, London RA (1987) Circ Res 60: 700
34. Margan E, Kitabaze M, Koretsume Y, Yue DT, Chacko VP, Pike MM (1990) Circ Res 66: 1255
35. Marban E, Kitakaze M, Chacko VP, Pike MM (1988) Circ Res 63: 673
36. Moon RB, Richards JH (1973) J Biol Chem 248: 7276
37. Jacobus WE, Press IH, Lucas SK, Kallman CH, Weisfeldt ML, Flaherty JT (1982) In: Nuccitelli R, Deamer DW (eds) Intracellular pH: its measurement, regulation and utilization in cellular functions. Alan R. Liss, New York, p 537
38. Brindle KB, Radda GK, Rajagopalan B (1989) Circulation 76: IV–341
39. Fisher MJ, Dillon PF (1987) Circ Res 60: 472
40. Meyer RA, Brown TR, Kushmerick MJ (1985) Amer J Physiol 248: C279
41. Clarke K, Balschi JA, Neubauer S, Kleber AG, Springer CS Jr, Smith TW, Ingwall JS (1988) Proceedings of the Eighth Annual Meeting of the Society of Magnetic Resonance in Medicine, p 665
42. Anderson RE, Nedelec J-F, Mills PA, Ingwall JS, Clarke K (1990) Proceedings of the Ninth Annual Meeting of the Society of Magnetic Resonance in Medicine, p 1243
43. Gupta RK, Bonenic JL, Rose ZB (1978) J Biol Chem 253: 6172
44. Gupta RK, Moore RD (1980) J Biol Chem 255: 3987
45. Headrick JP, Willis RJ (1989) Magn Reson Med 12: 328
46. Murphy E, Steenbergen C, Levy L, Raju B, London R (1988) J Mol Cell Cardiol 20 (Suppl III): 4
47. Chu SCK, Xu Y, Balschi JA, Springer CS Jr (1990) Magn Reson Med 13: 239
48. Kohler SJ, Perry SB, Stewart LC, Atkinson DK, Clarke K, Ingwall JS (1991) Magn Reson Med 18: 15
49. Naritomi H, Kanashiro M, Sasaki M, Kurbibayashi Y, Sawada T (1987) Biophys J 52: 611
50. Auffermann W, Camacho A, Wu S, Litt L, Parmley WW, Higgins CB, Wikman-Coffelt J (1987) Magn Reson Med 19: 1011
51. Balschi JA, Hohler SJ, Bittl JA, Springer CS Jr, Ingwall JS (1989) J Magn Reson 83: 138
52. Buster DC, Castro MMCA, Geraldes CFGC, Malloy CR, Sherry AD, Siemers TC (1990) Magn Reson Med 15: 25

53. Albert MS, Lee J-H, Springer CS Jr (1990) Proceedings of the Ninth Annual Meeting of the Society of Magnetic Resonance in Medicine, p 1269
54. Malloy CR, Buster DG, Castro MMCA, Geraldes CFGC, Jeffrey FMH, Sherry AD (1990) Magn Reson Med 15: 33
55. Nuebauer S, Balschi JA, Springer Cs Jr, Smith TW, Ingwall JS (1987) Circulation 76: IV–56
56. Albert MS, Lee J-H, Patlak CS, Fensternmacher JD, Springer CS Jr (1990) Proceedings of the Ninth Annual Meeting of the Society of Magnetic Resonance in Medicine, p 36
57. Duling BR (1983) In: Berne RM, Levy MM (eds) Physiology. CV Mosby, St Louis, p 824
58. LeFurgey A, Ingram P, Lieberman M (1988) Cell Calcium 9: 219
59. Balschi JA, Clarke K, Fetters JC, Springer CS Jr, Smith TW, Ingwall JS (submitted for publication)
60. Bernard M, Kohler SJ, Ingwall JS (unpublished results)
61. Linginger MI, Heigenhauset GJF (1987) J Appl Physiol 63: 426
62. Pike MM, Clarke D, Kitabaze M, Golden L, Madden MC, Cragoe E, Pohost GM (1990) Circulation 82: III–686

# In Vivo Phosphorus-31 NMR: Potential and Limitations

**M. Rudin and A. Sauter**

Preclinical Research, Sandoz Pharma Ltd., CH-4002 Basle, Switzerland

## Table of Contents



A number of endogenous compounds of biological significance contain one or more phosphate groups and are directly measurable in-vivo by $^{31}P$ NMR. The state-of-the-art of what and how information may be extracted from $^{31}P$ spectra and how such results may be interpreted, is reviewed.

The usefulness and potential of in vivo $^{31}P$ NMR to gain novel insights into the biology of whole organs and organisms critically depends on experimental aspects. Therefore, localization techniques, radiofrequency coils, relaxation times and sensitivity optimization, as well as editing techniques, such as broadband proton decoupling/nuclear Overhauser effect, editing based on $T_2$ effects, $^{1}H$–$^{31}P$ correlated 2D spectroscopy and magnetization transfer are discussed first.

In addition to the resonances of high energy phosphates (HEPs, i.e., phosphocreatine and ATP) and inorganic phosphate, which provide information on the cellular energy metabolism, signals of phosphomono- and diesters may be used as indicators of phospholipid metabolism, phosphatidylinositol turnover, glucose utilization etc. Moreover, intracelluar pH, free magnesium levels, intra- and extracellular volumes may be deduced from $^{31}P$ spectra.

Finally, the article focusses on the critical biological interpretation of results obtained by $^{31}P$ MRS and putative novel applications, both in experimental animal research and in clinical diagnosis.

# 1 Introduction

Among the nuclei studied in vivo by NMR, phosphorus ($^{31}P$) is by far the most prominent, due to several reasons: (1) $^{31}P$ is an NMR active nucleus with a relatively good sensitivity (7% of that of protons) and occurs at 100% natural abundance. (2) Some biological compounds of fundamental functional and structural importance contain one or more phosphate groups and are present in biological tissues in millimolar concentration, allowing their in situ determination by $^{31}P$ magnetic resonance spectroscopy (NMR) with acceptable precision ($\pm$ 10%), and sufficient temporal (minutes) and spatial ($cm^3$) resolution. (3) In addition, biologically relevant information such as tissue pH and free Mg levels can be derived from $^{31}P$ spectra. (4) The resonance frequencies of the $^{31}P$ containing compounds of biological interest are usually well resolved and easily identifiable (see Fig. 1) (5) In contrast to proton ($^1H$) NMR, where the dominant water signal has to be suppressed with suitable excitation pulses in order to make the compounds of biological interest detectable, no such dynamic range problem exists with $^{31}P$ NMR. A representative $^{31}P$ spectrum recorded from a normal rat brain is shown in Fig. 1 (compare Sect. 4.1 for the assignment of resonances).

In vivo $^{31}P$ NMR was used initially to study isolated cells [1]. The first $^{31}P$ spectrum recorded from an isolated organ was published in 1974 [2]. In the following years, $^{31}P$ NMR studies were carried out an isolated tissues and

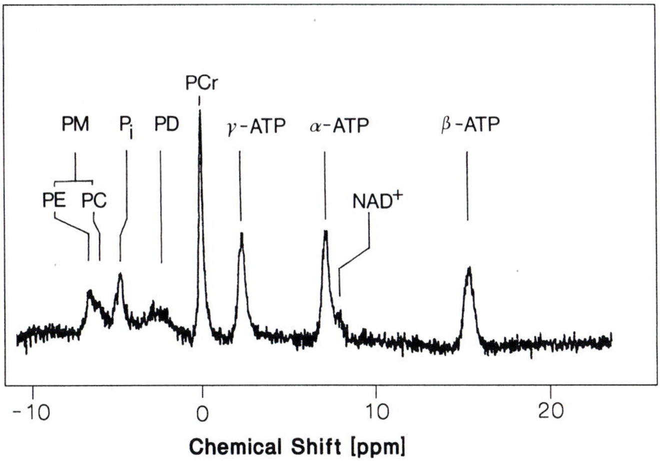


**Fig. 1.** Rat brain $^{31}P$ spectrum. The spectrum was recorded from a normal rat at 4.7 T using a 10 mm x 17 mm elliptical surface coil. The resonances of phosphomonoesters (PM), i.e., phosphorylcholine (PC) and -ethanolamine (PE), inorganic phosphate ($P_i$), phosphodiesters (PD) including phospholipids, phosphocreatine (PCr), the three resonances of ATP, and $NAD^+$/NADH may be identified. The chemical shift difference between the PCr and the $P_i$ signal, or alternatively between the $\gamma$- and $\alpha$-ATP resonance [47], depends the intracellular pH value

organs mainly using high-resolution spectrometers. The first localized spectra of an intact, living animal were obtained by magnetic field profiling (so-called 'topical NMR') [3]. Alternatively, surface coils, introduced by Ackerman et al. [4], provided a simple method to select superficially located regions of interest (ROI's) with good NMR sensitivity. Using this method, a large number of $^{31}P$ studies have since been published, mainly dealing with the energy metabolism under different pathophysiological and pharmacological conditions in various target tissues, such as brain and skeletal muscle. The non-invasive study of the deeper lying organs requires more sophisticated localization techniques such as ISIS [5], spectroscopic imaging [6–8] or DRESS [9].

The main focus of this chapter is on the potentials and limitations of in vivo $^{31}P$ NMR, a field that has started to explode in the past 15 years. Experimental aspects, such as localization techniques, probehead design, quantitation of spectral content etc., will be discussed only briefly, as these topics are extensively treated in other chapters of this series. Instead, we will discuss in detail, which parameters can be measured with $^{31}P$ NMR and which biologically relevant information can be deduced from these parameters.

## 2 Experimental Aspects

### *2.1 Localization*

The exact definition of the region-of-interest (ROI), i.e., localization and shape of the volume within the body from which NMR spectra are recorded, is a prerequisite for correct biological interpretation of the spectroscopic data. As mentioned above, the most simple technique is the use of surface coils [4], while more sophisticated techniques, such as ISIS [5] and variations thereof [10], DRESS [9] and spectroscopic imaging [6–8, 11, 12] rely on imaging principles. An alternative, though rather invasive technique to select deeper lying ROI is to chronically implant a NMR coil into the desired position within the body [13].

An important prerequisite to obtain high quality spectra is the good homogeneity of the magnetic field over the whole ROI. Commonly, the water signal is utilized for homogeneity optimization. When using surface coils, a better sensitivity of the proton channel is achieved by inserting an additional tuning/matching circuit in order to switch from $^{31}P$ to $^{1}H$ resonance conditions [14]. This additional coupling device is removed during collection of the $^{31}P$ data.

### *2.2 Radiofrequency Coils*

The large majority of spectroscopic investigations today is carried out using surface coils for radiofrequency (RF) transmitting and receiving [4]. The obvious advantages are the experimental simplicity and optimal filling factor, resulting in good sensitivity. The selected ROI depends on the size and the shape of the coil (see Bosch and Ackerman, Vol. 27 of this series). For a circular coil it corresponds roughly to a half sphere with the radius of the coil. Surface coils may easily be adjusted to the geometry of the organ of interest [15]. The inhomogeneous RF field inherent to surface coils used as transmitter devices give rise to problems such as inhomogeneous excitation and differential saturation in the ROI. These problems can be minimized by using adiabatic RF pulses [16, 17]. Extension of a simple single-frequency to the multiple-frequency tuned coil is straight forward (see Schnall, Vol. 26 of this series), allowing the simultaneous measurement of e.g., $^{1}H$ and $^{31}P$ spectra.

The most significant drawback of the surface coil is its limitation to superficial structures. Investigations of deeper lying organs may require whole body probes (see Link, Vol. 26 of this series). Typical structures are the Alderman–Grant type resonator [18–20], the birdcage probe [21], or the simple saddle coil. Such probes have to be double tuned or are double resonators ($^{1}H/^{31}P$), since the exact selection of the ROI relies on a $^{1}H$ image recorded beforehand.

## *2.3 Relaxation Times*

Exact knowledge of the different relaxation times of the substances of interest in various tissues is important for planning and optimizing NMR experiments. $T_1$ values of several phosphorus containing metabolites of biological interest have been determined by Eveloch et al. [22] and are given in Table 1. It may be noted that $T_1$, both in muscle and brain, significantly decrease as the magnetic field strength increases. This decrease may indicate an increasing contribution of the chemical shift anisotropy to relaxation, assuming the case of extreme narrowing with respect to molecular motions (fast correlation time limit) [22]. However, observed deviations from the $1/T_1 \alpha B_0^2$ dependence indicates that other mechanisms also influence the longitudinal relaxation time $T_1$. Measurements of the nuclear overhauser enhancement (NOE) demonstrated the importance of $^{1}H$–$^{31}P$ dipole–dipole interactions for the $^{31}P$ nuclear relaxation. Table 2 lists the relative NOE enhancements (with respect to the maximum NOE enhancement of $\eta_{max} = 0.5^{*}\gamma_H/\gamma_P = 1.24$) measured for adenosin triphosphate (ATP) phosphocreatine (PCr) and inorganic phosphate ($P_i$) in $H_2O$. These values together with the observation of almost zero enhancement in $D_2O$ show the importance of dipolar interaction between solute phosphate and solvent proton [23]. The smaller NOE for ATP and the different $\eta_{obs}/\eta_{max}$ values for its three phosphate groups may be attributed to the complexation with magnesium [23]. Table 2 shows that chemical shift anisotropy is the dominant relaxation process for ATP, while there is a significant dipolar contribution to the relaxation of PCr and $P_i$. The very low $T_1$ values observed in liver are due to the high concentration of paramagnetic ions in this organ, providing an additional and very efficient relaxation pathway. Similarly, the reduced $T_1$ values of the $\alpha$-, $\beta$-, and $\gamma$-phosphate

**Table 1.** Spin-lattice relaxation times $T_1$ (in s) of $^{31}P$ containing metabolites in various tissues at different magnetic field strengths[a]

| Tissue | Metabolite | 8.5T | 4.7T | 1.9T | 1.5T |
|---|---|---|---|---|---|
| Brain | α-ATP | 0.9 ± 0.2 | 1.5 ± 0.2 | | |
| | β-ATP | 0.7 ± 0.1 | 1.6 ± 0.0 | | |
| | γ-ATP | 1.1 ± 0.2 | 2.0 ± 0.3 | | |
| | PCr | 2.8 ± 0.7 | 4.4 ± 0.4 | 3.9 ± 0.9 | |
| | $P_i$ | 3.2 ± 0.6 | 5.6 ± 2.8 | | |
| Muscle | α-ATP | 1.1 ± 0.2 | 2.0 ± 0.1 | | 4.2 ± 0.6 [26] |
| | β-ATP | 1.2 ± 0.1 | 2.0 ± 0.3 | | 4.1 ± 1.3 [26] |
| | γ-ATP | 1.7 ± 0.2 | 2.8 ± 0.3 | | 3.9 ± 1.3 [26] |
| | PCr | 2.7 ± 0.7 | 4.5 ± 0.5 | 5.6 ± 0.8 | 6.5 ± 0.7 [26] |
| | $P_i$ | | | | 4.7 ± 0.4 [26] |
| Liver | α-ATP | 0.4 ± 0.1 | 0.4 ± 0.1 | | 0.7 ± 0.1 [26] |
| | β-ATP | 0.3 ± 0.1 | 0.2 ± 0.0 | | 0.4 ± 0.1 [26] |
| | γ-ATP | 0.4 ± 0.0 | 0.3 ± 0.0 | | 0.4 ± 0.2 [26] |
| | $P_i$ | 1.0 ± 0.3 | 1.0 ± 0.1 | | 0.4 ± 0.1 [26] |
| | PM | 1.3 ± 0.3 | 1.3 ± 0.2 | | |
| | PD | | | | 1.4 ± 0.2 [26] |
| Model Solution[b] | α-ATP | | 2.5 ± 0.2 [23] | | |
| | β-ATP | | 1.6 ± 0.1 [23] | | |
| | γ-ATP | | 1.7 ± 0.1 [23] | | |
| | PCr | | 5.8 ± 0.2 [23] | | |
| | $P_i$ | | 11.9 ± 0.3 [23] | | |

[a] Values are from Ref. [22], unless cited explicitely
[b] Solution: ATP, AMP, PCr, and $P_i$: 15 mM, NaCl: 100 mM, $MgCl_2$: 15 mM, buffers: (except for $P_i$) PIPES and HEPES, pH 7.4

**Table 2.** Relative nuclear overhauser enhancement (NOE) $\eta_{obs}/\eta^{a}_{max}$ for ATP, PCr and $P_i$ in $H_2O$ solution[b]. In $D_2O$ NOE effects of $0 > \eta_{obs} > -0.028$ have been measured at 4.7 T (Values are taken from Ref. [23])

| Metabolite | | $\eta_{obs}/\eta_{max}$ [%] | |
|---|---|---|---|
| | | 4.7 T | 2.35 T |
| ATP | α | 12 | 41 |
| | β | 8 | 17 |
| | γ | 15 | 29 |
| PCr | | 36 | 64 |
| $P_i$ | | 37 | |

[a] $\eta_{max} = 0.5 * \gamma_H/\gamma_P$
[b] Solution: ATP, AMP, PCr, and $P_i$ 15 mM, NaCl: 100 mM, $MgCl_2$: 15 mM, buffers: (except for $P_i$) PIPES and HEPES, pH 7.4

of ATP in brain, as compared to skeletal muscle, is indicative of a higher concentration of paramagnetic ions in this organ [22].

In contrast to the $T_1$ relaxation times, the *relative* line widths (in ppm) of the $P_i$, PCr and ATP resonances, are not dependent on the magnetic field strength. Hence $1/T_2^*$ is approximately proportional to $B_0$, suggesting that variations in line widths are determined by the chemical environment producing different chemical shifts and/or the $B_0$ heterogeneity due to variations in the sample susceptibility [22]. This explains e.g., that ATP J-couplings may be readily resolved under in vivo conditions at low field strengths (1.5 T), but not at higher field strengths (e.g., 4.7 T).

Tissue specific phosphorus $T_2$ relaxation times are one or two orders of magnitude smaller than the corresponding proton relaxation times. This has important practical consequences: Any pulse sequence containing periods of several ms with transverse magnetization (e.g., spin echo or stimulated echoes) is not practicable for $^{31}P$ NMR because of rapid signal dephasing. Therefore, $^{31}P$ localization techniques have to be used, in which the magnetization is oriented along the static field axis during waiting periods [5, 25] (see Ordidge and Helpern, Vol. 27 of this series).[a]

## 2.4 *Sensitivity Optimization*

Sensitivity, i.e. the signal/noise ratio (SNR), is a function of the specific relaxation parameters $T_1$ and $T_2^*$ (see Table 1), the magnetic field strength $B_0$, the RF pulse angle $\alpha$, the repetition delay $T_R$ and the total measurement time. The optimal pulse angle is defined by the relation $\cos\alpha = \exp(-T_R/T_1)$, with $\alpha = \gamma B_1 t_p$, $t_P$ being the pulse duration. Assuming a homogenous RF field and the sample to be the dominant noise source, the overall SNR for a given total measurement time is given by [22, 25],

$$\mathrm{SNR} = \frac{B_0 (T_2^*)^{1/2} \sin\alpha}{T_R^{1/2}} \cdot \frac{[1 - \exp(-T_R/T_1)]}{[1 - \exp(-T_R/T_1)\cos\alpha]}, \tag{1}$$

the second term being the saturation factor. In practice, the SNR can be optimized only by varying $T_R$ and adjusting the RF pulse angle accordingly. As the $T_1$ and $T_2^*$ values of each resonance depend on the chemical micro-environment, an individual, tissue specific optimal $T_R$ and pulse angle exist for each resonance. The above relation only holds for homogeneous RF excitation throughout the sample. For surface coils the pulse angle distribution over the ROI has to be considered.

As far as sensitivity is concerned, it would be most efficient to use very short recycle delays. The lower limit is given by the acquisition time needed for the data collection. However, as $T_R$ decreases, saturation effects become increa-

[a] Recently, Schenker et al. [146] have demonstrated the feasibility of spin echo sequences with echo times below 9 ms to obtain localized $^{31}P$ spectra of phantoms and human forearm. There are, however, considerable distortions as far as intensities (ATP) are concerned

singly important, resulting in significant intensity distortions for signals with different $T_1$'s. This is less of a problem if only the changes of a particular signal relative to its initial value are of interest. However, if different resonances have to be compared, correction with saturation factors, which account for the incomplete relaxation, are necessary.

## 3 Spectral Editing

### *3.1 Broadband Proton Decoupling/Nuclear Overhauser Effect*

At a low magnetic field strength (< 1.5 T) a residual water line width of 0.1 to 0.2 ppm corresponds to inhomogeneous line broadening of 3 to 5 Hz in the $^{31}P$ spectrum. Under this condition, $^{1}H$–$^{31}P$ scalar coupling could be the dominant factor for the experimentally observed line width, and proton decoupling should improve both sensitivity and resolution due to the collapse of multiplet structure and/or NOE.

In vivo proton decoupled $^{31}P$ spectroscopy [denoted as $^{31}P$-($^{1}H$) in the following] have been reported by several authors [27–31]. The main advantage of $^{31}P$-($^{1}H$) experiments is the increased information content due to the improved resolution and sensitivity. This especially holds true for the phosphomono- and diester region, where well resolved resonances of phosphorylcholine (PC), phosphorylethanolamine (PE), glycerophosphorylcholine (GPC) and glycerophosphorylethanolamine (GPE) are obtained. These compounds provide useful information on the turnover of cell membranes. In addition, the signal of PC may be used as a chemical shift reference since it shows little pH dependence (see Sect. 4.2). The extent of NOE enhancement, which depends on the strength of the dipolar interaction between the coupled spins, contains structural and motional information on the metabolites.

Unfortunately, the efficacy of $^{1}H$ decoupling decreases with increasing field strength. If the line width is mainly governed by tissue susceptibility effects and hence proportional to $B_0$, then the ratio of the coupling constant J over the line width $\Gamma$ (in Hz) will decrease proportionally to $1/B_0$, making proton decoupling with increasing $B_0$ less effective.

Figure 2 shows the effect of proton decoupling on the $^{31}P$ spectrum of the human calf muscle [28]. The most striking result is the improved resolution and SNR of the GPC resonance. In addition, the $\alpha$-ATP–NAD complex is also better resolved. NOE enhancement leads to a general increase in sensitivity. Similar enhanced spectra were obtained from the human brain, liver, and heart [28]. In the heart, decoupling resulted in an acceptable resolution of the 2,3-diphosphoglycerate (2,3-DPG) and $P_i$ signals.

An important aspect in such studies, especially when involving humans, concerns the RF power deposition in the tissue. The FDA guidelines for the

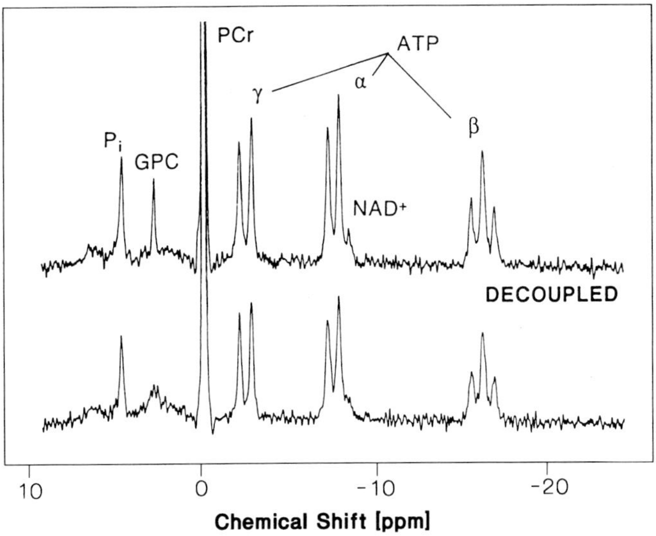


**Fig. 2.** Effects of proton decoupling on $^{31}P$ spectrum of human calf muscle. Spectra were recorded at 1.5 T using a 14 cm diameter surface coil. The lower trace represents the conventional spectrum. Proton decoupling (applying a train of inversion pulses through a seperate $^{1}H$ coil) leads to a general increase in signal-to-noise ratio due to nuclear Overhauser enhancement (see Table 2). The decoupled spectrum shows a well resolved resonance in the PD region attributed to glycerophosphoryl-choline (GPC). In addition the peak of $NAD^+$ is somewhat better resolved (adapted from Ref. [28], reproduced with permission)

specific tissue absorption rate is 3.2 W/kg. Efficient decoupling using WALTZ sequences [32] may be achieved well within these limits [27, 28]. The use of adiabatic pulses (see Garwood and Ugurbil, Vol. 26 of this series) allows homogeneous decoupling conditions to be maintained throughout the ROI, even with surface coils.

## 3.2 *Editing Based on $T_2$ Effects*

The large difference in $T_2$ values observed for the $^{31}P$ containing metabolites (e.g., $\beta$-ATP ~ 10 ms, PCr ~ 190 ms [34]) may be exploited for spectral editing e.g., in a spin-echo experiment [33]. For conventional spectroscopy this is not of importance, since sufficient spectral resolution is obtained with standard spectroscopic techniques. However, when carrying out chemical shift imaging experiments, spin echo measurements may avoid the problems of spectral deconvolution. Hsieh and Balaban [35] and Mora et al. [36] have used spin echo $^{31}P$ imaging for mapping of the PCr signal with an without saturation of the $\beta$-ATP resonance in order to determine the spatial variation of the creatine kinase forward rate constant.

## 3.3 Removal of the 'Broad Hump' in Brain Spectra

A problem encountered in $^{31}P$ spectroscopy of tissue (especially brain) is the presence of a broad spectral component ('broad hump'), which may originate from signals due to cranial bone and membrane phospholipids with restricted motional freedom [34, 37–39]. This 'broad hump' interferes with the accurate quantitation of the superimposed resonances and therefore, is commonly removed by various data post-processing methods, such as the convolution difference technique [40], removal of the first few data points of the FID, operator defined baseline correction, etc. These manipulations may, however, also introduce new artefacts leading to incorrect quantitations.

An alternative approach is to experimentally remove the 'broad hump' e.g. by doing a spin-echo experiment, exploiting the $T_2$ difference between the broad component ($\sim 0.2$ ms) and the resonances of interest (e.g. $T_{2,\beta-ATP} \sim 10$ ms) [41, 42]. This method works reasonably well for brain, although some intensity distortion occurs due to the $T_2$ differences which also exists among the resonances of interest. Another method is based on the assumption that the 'broad hump' arises from homogeneous broadening. It may therefore be suppressed by selective irradiation, using low power continuous wave irradiation via a second RF channel, or a DANTE pulse train [43]. Ackerman et al. [34] and Thoma et al. [44] have shown that this yields an effective saturation of the undesired resonance.

## 3.4 $^1H$–$^{31}P$ Correlated 2D Spectroscopy

The assignment of $^{31}P$ resonances in tissues is usually not always straightforward because their chemical shifts depend on the environment. An immediate correlation between the results obtained with extracts under high resolution conditions and those in vivo may not be obvious. Moreover, the spectral resolution in vivo may be insufficient to clearly identify resonances and determine their frequency with sufficient accuracy. This holds true, in particular, for the phosphomono- and diester region where only poorly resolved broad structures can be observed under most circumstances. $^1H$–$^{31}P$ correlated spectroscopy is an effective tool for the analysis of poorly resolved spectral regions [45] (see Navon et al., Vol. 27 of this series).

## 3.5 Magnetization Transfer

Magnetization transfer experiments reveal connections among different compounds as they occur via chemical exchange, cross-polarization, transport processes etc. The main application of magnetization transfer measurements in $^{31}P$ spectroscopy is the measurement of rate constants and fluxes through enzymatic pathways, such as the creatine kinase reaction, ATP synthase, adenylate kinase, etc. Magnetization transfer is treated in detail in another chapter of this series (Rudin and Sauter, Vol. 27).

## 4 Information Content of $^{31}$P Spectra

### 4.1 Phosphorus Containing Molecules Assessable by NMR

Figure 1 shows a typical $^{31}$P spectrum of rat brain. The resonances of inorganic phosphate ($P_i$), phosphomonoesters (PM), phosphodiesters (PD), phosphocreatine (PCr), adenosin triphosphate (ATP; $\alpha$-, $\beta$-, and $\gamma$-resonance) and nicotinamide dinucleotide (NAD) can be identified. Adenosin diphosphate (ADP) is not visible under physiological conditions, since only a small fraction of the total cellular ADP is in the cytosol. The major fraction is bound to macromolecules and hence immobile, leading to broad, undetectable signals. The ADP concentration may be derived via the creatine kinase equilibrium constant, provided the total creatine content of the tissue is known (see below). The PM signal contains the resonances of phosphatidylcholine (PC),-ethanolamine (PE) and sugar phosphates, while the PD peak is constituted of glycerophosphorylcholine (GPC) and -ethanolamine (GPE). Only the $\beta$-resonance is a pure nucleotide-triphosphate (NTP, mainly ATP) signal, the $\gamma$-peak contains contribution from NDP's and the $\alpha$-signal both from NDP's and NAD/NADH.

The assignment of recorded in-vivo resonances simply by comparison with tissue extracts is problematic, since the resonance frequencies of several metabolites depend critically on the tissue specific chemical environment, such as pH, free magnesium concentration etc. On the other hand, this dependence can be exploited for the non-invasive determination of the respective parameters.

### 4.2 Determination of Intracellular pH Value

In general, the resonance frequency of a nucleus in a particular molecule is influenced by any changes in structure, including e.g., protonation, so that the protonated form HA has a frequency different from that of the unprotonated molecule A. In particular, the resonance frequencies of phosphate containing compounds greatly depend on the extent of protonation and hence, on the pH of the environment. At a physiological pH of around 7 e.g., about half of the inorganic phosphate $P_i$ occurs as $H_2PO_4^-$, the other half as $HPO_4^{2-}$. However, even though the resonance frequencies of the two species differ, only a single $P_i$ frequency is observed in a $^{31}$P spectrum. The exchange between the protonated and deprotonated form, $HA \rightleftharpoons H^+ + A$ is rapid compared to the difference in resonance frequencies $|f_{HA} - f_A|$ [46]. The time-averaged, single resonance frequency $\delta$ of the two species occurs at

$$\delta = x_A \delta_a + (1 - x_A)\,\delta_{HA}, \tag{2}$$

where $x_A$ indicates the molar fraction of the deprotonated form. Combining this equation with the well known relationship $pH = pK_A + \log(x_A/(1 - x_A)$ leads to:

$$pH = pK_A + \log\frac{\delta - \delta_{AH}}{\delta_A - \delta}. \tag{3}$$

For a precise pH measurement by NMR, the following criteria have to be fulfilled: (1) $^{31}P$ resonances used for pH determination should be readily detectable under all physiological conditions. (2) The resonance frequencies should be unaffected or predictably affected by other parameters such as the $Mg^{2+}$ concentration. (3) One resonance frequency or chemical shift difference has to vary in a sensitive way in the pH range of interest, i.e. for most biological applications pH 6.5 to 7.5. Many of the resonances in the $^{31}P$ spectrum have pK values around 7 and may, therefore, be used as physiological pH indicators. A comprehensive list of pK values for several $^{31}P$ resonances is given in ref. [47]. It is important to realize that the pK and hence $\delta_A$ and $\delta_{AH}$ depend on the ionic strength of the medium in which they are measured. The intrinsic reference for the chemical (i.e. pH) shift of the former, should ideally not depend on variations of the cellular milieu.

In in vivo studies, the chemical shift of the $P_i$ signal relative to the PCr signal is used as a pH probe. A selection of published $pK_a$, $\delta_a$, and $\delta_b$ values is given in Table 3. However, there are several limitations in using of the $P_i$–PCr chemical shift difference as a pH indicator: (1) Some tissues such as liver, kidney, or certain tumors do not contain PCr. (2) In pathological situations, such as hypoxia or ischemia, the PCr signal may vanish completely, preventing an accurate determination of the pH value. (3) Alternatively, in normal tissue the $P_i$ signal may be too small for detection. (4) In certain organs, such as the heart, the $P_i$ signal may overlap with the 2,3-diphosphoglycerate (2,3-DGP) resonance and an accurate determination of the $P_i$ resonance position is not possible. If these caveats

**Table 3.** Determination of pH from the chemical shift difference $\delta$ of the $P_i$ and PCr resonance; experimentally derived parameters of the Henderson–Hasselbach equation $pH = pK_a + \log[(\delta - \delta_a)/(\delta_b - \delta)]$

| $pK_a$ | $\delta_a$ | $\delta_b$ | Reference |
|---|---|---|---|
| 6.84 | 3.22 | 5.73 | 2 |
| 6.72 | 3.27 | 5.69 | 48 |
| 6.75 | 3.15 | 5.65 | 49 |
| 6.66 | 3.08 | 5.57 | 50 |
| 6.77 | 3.22 | 5.73 | 51 |
| 6.90 | 3.29 | 5.80 | 52 |
| 6.75 | 3.29 | 5.70 | 53 |
| 6.79 | 3.25 | 5.75 | 54 |
| 6.77 | 3.29 | 5.68 | 55 |

can be neglected, the accuracy achieved is typically 0.05 pH units. The pH values obtained from $^{31}P$ NMR essentially represent the intracellular, cytoplasmatic pH [48].

The difference of the $\gamma$-ATP and $\alpha$-ATP has also been proposed as a pH probe [47, 56]. Like PCr, the $\alpha$-ATP resonance shows little pH dependence in the physiological pH range and can be used as a reference, while the $\gamma$-ATP resonance varies between $\delta = -9.107$ ppm for the protonated and $-4.777$ ppm for the deprotonated form [47]. Unfortunately, the resonance frequencies of ATP depend markedly on the intracellular concentration of multivalent cations, such as $Mg^{2+}$. In fact, the chemical shift difference between the $\gamma$-ATP and $\beta$-ATP resonance can used as an indicator for the intracellular free $Mg^{2+}$ concentration [57, 58]. Therefore, accurate pH determinations are only possible with known $Mg^{2+}$ concentrations.

Finally it is possible to introduce exogenous $^{31}P$ containing pH probes. Shulman et al. [59] analyzed the chemical shift of fructose-1-phosphate (fructose-1-P), formed in vivo from exogenously administered fructose. However, the pK value of fructose-1-P is pK = 6.1, which is at the edge of the physiologically useful range. Moreover, a fructose load is accompanied by the formation of inositol monophosphate, the signal of which may interfere with that of fructose-1-P, preventing an accurate measurement of its frequency. Alternatively, the resonance position of the 2-deoxyglucose-6-phosphate (DOG-6-P) signal, observed after administration of DOG, has been used to derive pH values [48, 60]. This procedure might have its merit if at the same time glucose utilization is measured, by monitoring the formation of DOG-6P after DOG administration [61, 62].

An attractive application of exogenous pH indicators which do not cross the cellular membrane has been reported by Clarke et al. [63]. By adding phenylphosphonic acid (PPA) to the perfusion medium, these authors have measured simultaneously the intra- and extracellular pH in perfused rat hearts by analyzing the chemical shifts of the $P_i$ and PPA signals, respectively. They were thus able to follow the sacrolemmal pH gradient under normal and pathologic conditions.

### *4.3 Determination of Intracellular Free Magnesium*

Coan and Hughes [64] were the first to show that the resonance frequencies of the $\beta$- and $\gamma$-phosphate, but not that of the $\alpha$-phosphate of ATP depend on the $Mg^{2+}$ concentration in the medium, indicating that this cation primarily binds to these two phosphate groups. Gupta et al. [65, 67] suggested to use the chemical shift difference between the $\alpha$- and $\beta$- or $\gamma$-resonance as a measure for the free $Mg^{2+}$ concentration. Assuming that the exchange $Mg^{2+} + ATP \rightleftharpoons (Mg^{2+}\ ATP)$ is rapid as compared to the respective chemical shift difference, only a single, time-average signal is observed with a chemical shift of

$$\delta_i = x_f \delta_{i-ATP} + (1 - x_f)\, \delta_{i-MgATP}, \tag{4}$$

where i stands for $\alpha, \beta$, and $x_f$ for the molar fraction of free ATP. Introduction of the dissociation constant $K_{D,MgATP} = [Mg^{2+}][ATP]/[MgATP]$ yields for the free $Mg^{2+}$ concentration (following the notation of Gupta [57]).

$$[Mg^{2+}] = K_{D,MgATP}(x_f^{-1} - 1). \tag{5}$$

In terms of chemical shift difference $\delta_{ij} = \delta_i - \delta_j$, this leads to

$$x_f = (\delta_{ij} - \delta_{ij,MgATP})/(\delta_{ij,ATP} - \delta_{ij,MgATP}), \tag{6}$$

allowing the calculation of $[Mg^{2+}]$ upon insertion into Eq. (5).

Two aspects, which are of importance for a correct interpretation of the data, should be mentioned: (1) Despite the fact that ATP is present in various subcellular compartments with probably different $Mg^{2+}$ levels, only one resonance for each phosphate group is observed. This means that the amount of ATP other than cytosolic is either negligible or largely in an immobilized form, so that these signals are broadened beyond detection. Alternatively, one may envisage a rapid exchange between all compartments, leading to single, averaged ATP signals. In the first two cases, the measurements would represent the cytosolic free magnesium, while in the last case the average of the free $Mg^{2+}$ concentration of all compartments involved in the exchange would be determined. (2) The resonance frequencies of ATP are also affected by the intracellular pH value as discussed above [47]. In order to reliably measure the $Mg^{2+}$ concentration, we have to make sure that the pH remains unchanged or changes in a well defined way. Vink et al. [68–69] have taken this aspect into account by using a pH correction for the MgATP dissociation constant.

A completely different approach to assess the intracellular free magnesium has been proposed by Cohen and Burt [70], who have shown for perfused frog gastrocnemius muscle at 4 °C that the transverse relaxation time ($T_2$) of PCr depends on the $Mg^{2+}$ concentration in the perfusate. This technique, however, is not applicable to in vivo studies in intact animals (humans), since the $T_2$ values are considerably affected by dynamic processes as well (e.g., creatine kinase exchange).

## *4.4 Determination of Intra- and Extracellular Volumes*

In order to make accurate statements about concentrations, the exact volumes of the different compartments (e.g., intra- and extracellular) have to be known. Clarke et al. [71] have assessed the total and extracellular water volumes in perfused rat hearts by using phosphonated tracer compounds: Phenylphosphonic acid (PPA), which is unable to cross the cell membrane, is used as a marker for the extracellular space, dimethylmethylphosphonate (DMMP) for the total water space. These two compounds neither affect the myocardial contractile function nor the high energy phosphate levels. A glass capillary adjacent to the heart, containing 2-aminoethylphosphonate (2 AEP), may serve as an external concentration reference. The volumes are then calculated according to

$$V_{total} = \frac{m_{2AEP}}{S_{2AEP}} \cdot \frac{S_{DMMP}}{c_{DMMP}}, \tag{7}$$

and

$$V_{extra} = \frac{m_{2AEP}}{S_{2AEP}} \cdot \frac{S_{PPA}}{c_{PPA}}, \tag{9}$$

where $m_{2AEP}$ referes to the total amount of 2 AEP in μmol in the capillary. $S_i$ represents the peak area of compound i in the fully relaxed spectrum, and $c_i$ its concentration in μ mol/ml in the perfusate. The intracellular volume is then obtained as $V_{intra} = V_{total} - V_{extra}$, allowing the calculation of the tissue concentration e.g., of ATP according to

$$c_{ATP} = \frac{m_{2AEP}}{S_{2AEP}} \cdot \frac{S_{\beta-ATP}}{V_{intra}}. \tag{9}$$

## 5 In Vivo $^{31}$P Spectroscopy: Selected Examples

### *5.1 High Energy Phosphate Metabolism*

The metabolism of high energy phosphates (HEPs, i.e., ATP and PCr) has been studied with $^{31}$P NMR in different species and tissues, and under a variety of pathophysiological conditions. Many closely related studies have been published.

A simplified scheme describing the turnover, i.e. synthesis and utilization of ATP is depicted in Fig. 3. An important energy source required for ATP synthesis is glucose, which under normoxic conditions is completely oxidized, via glycolysis and oxidative phosphorylation, to $CO_2$ and water, yielding 38 moles of ATP per mole of glucose. Under hypoxic conditions, ATP synthesis from glucose is still possible, but very inefficient; the yield is only 2 moles ATP per mole of glucose, since the oxidation stops, after glycolysis, at the levels pyruvate ↔ lactate. With combined $^{1}$H/$^{31}$P spectroscopy, lactate levels may be determined by $^{1}$H NMR in parallel to the already mentioned molecules participating in HEP turnover, providing direct information on the anaerobic ATP synthesis. An important cellular energy reservoir, which can be tapped very rapidly, is provided by PCr, which via the creatine kinase reaction transfers its phosphate group to ADP. ATP is utilized, i.e., hydrolysed to ADP and $P_i$, by every cellular process requiring free energy, such as the biosynthesis of cell constituents from simple precursors, contraction and motility which are forms of mechanical work and active transport, which is a reflection of osmotic or concentration work. Hence, it should be possible to gain some fundamental

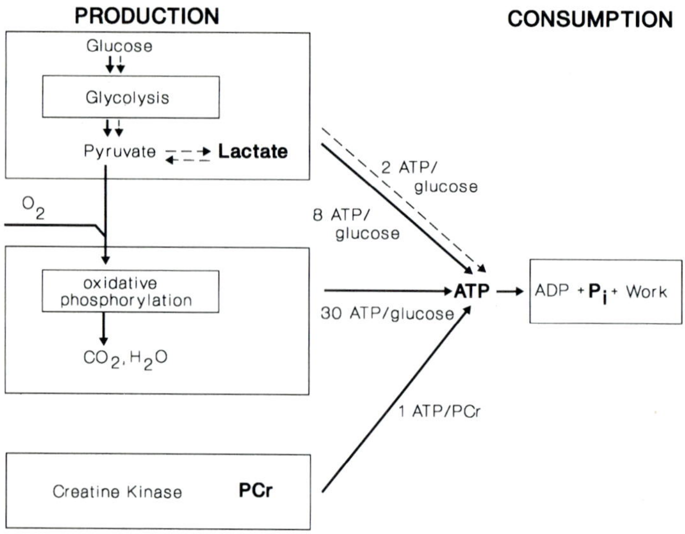


**Fig. 3.** Simplified scheme of high energy phosphate turnover. ATP is produced from glucose aerobically via glycolysis followed by mitochondiral oxidative phosphorylation (solid arrows), yielding a total of 38 moles ATP per mole glycose. In the absence of oxygen ATP is synthetized anaerobically (dashed arrows) with a yield of only 2 moles ATP per mole glucose. PCr represents an important energy resevoir, which may be activated very rapidly for ATP synthesis via the creatine kinase reaction. ATP is consumed by every biological process in the cell requiring energy. The metabolites printed in bold may be monitored in vivo using either $^{31}P$ or $^{1}H$ NMR

**Table 4.** Concentration of PCr, ATP, ADP, inorganic phosphate, and intracellular pH values for various tissues. Values are given in μ moles/g wet weight

| Tissue | PCr | ATP | ADP[a] | $P_i$ | pH | PCr/ATP | Ref. |
|---|---|---|---|---|---|---|---|
| Brain | $7.1 \pm 0.9$[c] | 3.0[b] | $7 \pm 1$ | | | $2.4 \pm .3$ | 73 |
| | $4.9 \pm 1.0$ | $2.5 \pm .5$ | | $2.0 \pm .5$ | | $2.0 \pm .6$ | 74 |
| | 5.0 | 3.0 | | | $7.1 \pm 0.1$ | 1.7 | 75 |
| Heart | $9.6 \pm 0.5$[c] | 5.0[b] | $12 \pm 3$[c] | | | $1.9 \pm .2$ | 73 |
| | $9.8 \pm 0.2$[c] | $5.3 \pm .1$[c] | | $1.7 \pm .3$[c] | $7.0 \pm 0.1$ | $1.8 \pm .1$ | 54 |
| | | | | | $7.15 \pm .03$ | | 71 |
| Muscle | $23.0 \pm 1.0$[c] | 6.0[c] | $20 \pm 4$[c] | | | $3.8 \pm .2$ | 73 |
| | 24.4[d] | 5.7[d] | 9.5[d] | 3.6[d] | 7.04 | $4.1 \pm .2$ | 76 |
| Liver | — | $2.7 \pm .3$ | | $1.8 \pm .3$ | 7.2 | — | 77 |
| Kidney | — | $2.1 \pm .2$ | | | | | 78 |
| | — | $4.7 \pm .8$ | | $2.4 \pm .3$ | | — | 79 |
| | — | $2.9 \pm .5$ | | | | — | 80 |

[a] Concentration in nmoles/g wet weight. The values are derived via the creatine kinase reaction using an equilibrium constant of $166 \times 10^9\ M^{-1}$ [73, 76]
[b] Values determined with classical biochemical analyses, taken as a reference for the NMR concentration determination
[c] Calculated from mM assuming of 0.65 of wet weight represents intracellular water [73]
[d] Values calculated using a ratio wet (weight/dry weight) of 4.35 [76]

insight into the normal and pathological functioning of organs by studying HEP metabolism.

Using NMR, *relative* tissue concentrations of HEPs can be determined quite easily. On the other hand, a variety of problems is associated with the determination of *absolute* concentrations [72] (see E. Cady, this series). Considerable discrepancies between concentrations determined by NMR and biochemical methods have been noted, e.g. NMR determinations of ADP yields μM concentration compared to mM determined biochemically. This is due to the fact that HEPs (and other metabolites) are known to be compartmentalized and only that the freely mobile phosphates, which constitute only a fraction of the total, biochemically detected HEP pool, is NMR visible. Table 4 lists absolute HEP concentrations and intracellular pH values for several tissues. Different levels of ATP and PCr, both absolute and relative, are found in different organs. In fact, the ratio PCr/ATP may be used as an index for tissue specificity and thus, as an indicator of the qualitity of volume selection in $^{31}$P NMR.

The tissue concentrations of HEPs depend on the rates of their synthesis and consumption. Pathologies involving impaired synthesis or excessive consumption are known in humans. HEP synthesis may be reduced either by insufficient oxygen (hypoxia) or glucose (hypoglycemia) supply, or by insufficient blood perfusion (ischemia). In the latter case, the tissue suffers not only from lack of oxygen and glucose or other substrates of HEP synthesis, but also from the insufficient removal of waste products e.g., lactate. Excessive consumption of HEPs, eventually leading to accumulation of lactate, occur under pathological conditions in the central nervous system e.g., during status epilepticus, but also in the periphery e.g., during muscular exertion resulting in myalgia. The organs which are most vulnerable to such pathologies are the brain and heart, and also skeletal muscle. Related disorders, which may be diagnosed and investigated biochemically and pharmacologically both in humans and animal models using $^{31}$P NMR, are stroke, head trauma, multi infarct dementia and other neurologic disorders associated with impaired blood flow [81–88] and/or hypoxia [90–96], epilepsy [97], myocardial infarction [98–100], intermittent claudication [101, 102], and muscular disorders [103, 104].

Impaired synthesis and/or excessive consumption of HEPs lead to typical changes in the $^{31}$P spectrum which are similar in all tissues. First, the PCr, later also the ATP resonances decrease, with a concomitant build-up of $P_i$. At a later stage the ATP concentration may also decrease. The rates of HEP disappearance and $P_i$ build-up vary with the type of tissue, reflecting organ specific differences in the capability to cope with a deranged HEP turnover. In brain e.g., the HEP levels are completely depleted within minutes after loss of aerobic glycolysis due to cardiac arrest [105], while in nonstimulated skeletal muscle the ATP levels remain well preserved for more than one hour [106]. An impaired aerobic HEP synthesis, which under sustained consumption leads to a lowering of tissue HEP levels, is in general compensated by an accelerated anaerobic glycolysis, producing lactate, which accumulates in the tissue and causes the pH to drop from 7.15 to below 6.5. Again, the rate and extent may differ with the tissue. The extent of lactate production depends mainly on the amount of glucose available for oxidation. In hyperglycemic animals with high blood glucose, more lactate is produced and the tissue concentrations are hence higher than in normo- or hypoglycemic animals, resulting in turn, in a lower pH [107]. In general, there is a good correlation between the lactate concentration and the intracellular pH. In brain, a correlation coefficient of $r = 0.95$ has been found [107]. This is, however, not always true. In gerbils, Allen et al. [108] have reported that lactate removal proceeds at a significantly slower rate than the normalization of intracellular pH. Obviously pH normalization occurs by processes not involving lactate. Moreover, lactate was observed in chronic cerebral infarction [109], despite the normal appearance of the $^{31}$P spectrum [110]. The overall intensity of the $^{31}$P signals in chronically infarcted brain regions is reduced compared to intact brain [86].

It has to be realized, however, that in general only drastic alterations of either HEP synthesis or consumption will lead to the typical changes described above. During mild to moderate hypoxia brain ATP levels remain essentially unchanged and only mild acidosis is observed [89–91, 94]. Similarly, cerebral blood flow may be reduced by more than 50% without noticeable effects on the level of ATP [88, 111]. Also increases or reductions of HEP consumption, elicited either by epileptogenic or anaesthetic drugs such as bicuculline or barbiturates, are not accompanied by corresponding changes in the $^{31}$P spectrum [89, 79, 112].

The finding that HEP levels remain remarkably constant, despite distinct variations in their synthesis or consumption, suggests that they are well regulated: increased consumption is matched by corresponding increases in synthesis and vice versa. This has been shown to be true experimentally in the isolated paced rat heart: variations of the workload had little or no effect on the HEP levels, while the oxygen consumption, which is a measure of aerobic HEP synthesis, correlated well with the heart rate-pressure product [116]. The same holds true also for other organs, e.g., the brain [112] (see below). It can therefore be concluded that the steady state tissue levels of HEPs per se are rather insensitive indicators of organ function.

A more sensitive parameter reflecting the functioning of an organ should be the turnover rate of HEP, which has to be tightly coupled to the energy consumption of the tissue, and which may be assessed in vivo in an elegant way using magnetization transfer techniques (see M. Rudin and A. Sauter, this volume). Alternatively, the turnover rate of HEP may be estimated by measuring the rate of disappearance of ATP and PCr, or alternatively, the build-up of $P_i$ following inhibition of aerobic HEP synthesis [89, 104, 105, 107]. Since HEP synthesis requires oxygen, a complete inhibition of oxygen delivery to an organ, for example by cardiac arrest, can be considered as an almost complete block of HEP synthesis, except for anaerobic glycolysis producing lactate. If anaerobic glycolysis is constant, any change in the rate of HEP disappearance following cardiac arrest may be interpreted as a change of HEP utilization. The extent of anaerobic glycolysis can also be estimated by in vivo NMR, either by pH measurement, which correlates with the tissue lactate levels, or by direct determination of tissue lactate levels by $^{1}$H-NMR [113, 114]. The approach of measuring HEP turnover by mointoring the adaptation of HEP levels to situations of impaired aerobic HEP synthesis has been frequently utilized to study energy metabolism in brain [81–89, 94, 105, 107], heart [98, 100] and skeletal muscle [103, 104].

Further insight into the turnover, specifically the synthesis rate of HEPs may be obtained by measuring the replenishment of depleted HEP tissue pools. Depletion is accomplished by forced energy consumption, e.g., electrical stimulation of calf or heart muscle contraction, or drastically decreased blood supply, or a combination of both. Following restoration of blood flow, the time constants for the normalization of HEP levels may be determined, which can be considered a parameter of the HEP synthesis rate [101, 102].

An alternative way to assess the energy consumption of an organ is to measure consumption of other substrates (i.e. $O_2$, glucose) utilized. Positron emission tomography in humans and autoradiographic methods [115] in animals are presently the established techniques for this purpose. Recently, it has been reported that the formation of phosphorylated sugars may be easily followed in vivo using $^{31}P$ spectroscopy [61, 62]. After peripheral administration of 2-deoxyglucose (DOG), Koga and Miura [62] monitored in rat brain, , the formation of DOG-6-phosphate at different stages of brain activity i.e., during barbiturate or nitrous oxide anaesthesia and bicuculline induced seizures. The accumulation of DOG-6-phosphate in brain is significantly increased during epileptic seizures, suggesting a higher glucose utilization and HEP synthesis rate, mandatory to keep-up with the higher cerebral HEP consumption, while the opposite is observed with barbiturates. In agreement with these results, we have observed that in rat brain the flux through the creatine kinase reaction, which has been shown to correlate in the glucose perfused rat heart with the workload, and $O_2$ consumption [116] is increased 2–3 fold during epileptic seizures induced by bicuculline and decreased following barbiturate administration [112].

In summary, $^{31}P$ NMR can be used to investigate the biochemistry and pharmacology of energy metabolism of tissues and organs noninvasively. However, the tissue levels of energy metabolites frequently are unchanged, because of efficient regulation mechanisms, and therefore are not sensitive indicators for subtle, but eventually relevant changes taking place in an organ. Instead, the direct or indirect assessment of HEP turnover rates is probably a more sensitive and relevant probe for the characterization of normal and pathological tissue states. $^{31}P$ NMR will certainly increase further our understanding and diagnosing capabilities of a number of human disorders related to energetics [104].

### *5.2 Phospholipid Metabolism*

The phosphomonoester (PM) resonances (5 to 8 ppm with respect to PCr chemical shift) comprise the signals from phosphoryl-choline (PC) and -ethanolamine

(PC) [117, 118]. The main constituents of the phosphodiester (PD) resonances (2 to 4 ppm with respect to PCr) are signals from glycerol-phosphodiesters, mainly glycerol-3-phosphorylethanolamine (GPE) and -choline (GPC), and a broad resonance from the phosphoryldiester linkeages within phosphorylated glycolipids and glycoproteins [39, 117, 119–121]. Thus, the PM and PD resonances may provide direct information on the synthesis and degradation of cellular membranes, as depicted in Fig. 4. Elevated PM and PD signals indicate increased turnover of membrane phospholipids. Well known examples are brain maturation [118, 112–124] and eventually degeneration [117], and the proliferation of tumors [125].

$^{31}$P spectra of neonates show a large PM signal, attributed to a high concentration of PE [118]. This is indicative of a high rate of phospholipid synthesis. During the first 6 months of postnatal life the ratio PM/PD, which reflects the ratio of phospholipid anabolic/catabolic activity [117], decreases to reach the ratio observed in adults [123, 126]. The myelination process, which may be monitored with $^{1}$H NMR, is not reflected by parallel changes in the $^{31}$P spectrum up to 4 years [128]. In fact, the most prominent changes in the $^{31}$P spectrum occur within the first 60 weeks of postconceptional life, while the most significant alterations in NMR images occur later [123]. Therefore, $^{1}$H NMR and $^{31}$P spectroscopy represent complementary tools to characterize brain maturation.

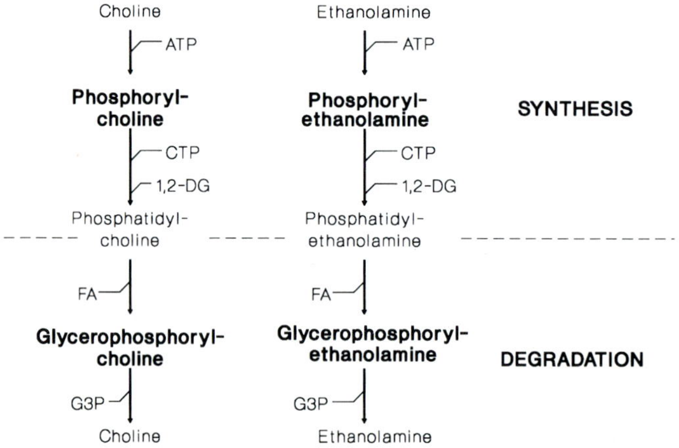


**Fig. 4.** Schematic representation of phospholipid turnover. The two membrane phospholipids phosphatidyl-choline and -ethanolamine are synthetized from cytosolic choline and ethanolamine. Phosphorylation catalyzed by the respective choline and ethanolamine kinases yields the NMR observable phosphorylcholine (PC) and -ethanolamine (PE), which via PC(PE)-CTP- and -1,2-DG-transferases form the respective phosphatidyl compounds. Degradation catalyzed by phospholipases and phosphodiesterases, leads to the NMR detectable glycerophosphoryl-choline (GPC) and ethanolamine (GPE), and finally choline and ethanolamine. The following abbreviations are used: ATP: adenosine triphosphate, CTP: cytidyltriphosphate, 1,2-DG: 1,2-diacylglycerol, FA: fatty acids, G3P: glycrol-3-phosphate

The ratio PM/PD has been suggested as an index of early brain maturation [123].

In aging rat brain (24 month old rats), both PM and PD signals are slightly increased, indicating an elevated turnover of membrane phospholipids. Similarly, elevated PM and PD level's have been observed in brain areas of patients with Alzheimer's disease (AD). The same areas were identified post-mortem as showing the typical neuropathological changes [117]. Even more intriguing is the finding that in regions, devoid of the neuropathological changes typical of AD, slightly increased PM levels are detected [117]. This may suggest that PM elevations in AD brains antedate cell morphological changes [117]. Moreover, it has been suggested that the correlation of $P_i$ with PM signal intensities allows the discrimination of multi-infarct dementia and AD [129]. At present, the above conclusion remain speculative however.

A large number of $^{31}$P spectroscopic studies deal with the metabolism of HEPs and phospholipids in neoplastic tissue [130–132]. Despite a large variation in spectral patterns published, there are some common features, such as an elevated PM resonance, reflecting an increased synthesis rate of membrane phospholipids in a proliferating tumor [125]. Eveloch and coworkers [125] have found that the PM/ATP ratio correlates with the size of adenocarcinoma tumors in mice. A similar correlation was observed in our lab: the PM/ATP ratio correlated with the size of dimethylbenzoanthracene induced mammary tumors in rats ($r = 0.83$, $N = 10$) [133]. Increased phospholipid metabolic activity in neoplastic tissue is also reflected by an increased choline peak, visible in proton spectra, as reported for several brain tumors [134] (Williams, this volume).

In summary, $^{31}$P NMR spectroscopy may provide information on phospholipid metabolism in vivo. Potential applications comprise the study of neurodegenerative diseases (AD, Huntington's chorea), brain maturation and tumor metabolism. A detailed interpretation of the spectra is hampered by the poor resolution in the range of the PM and PD signals. At lower field strengths ($< 1.5$ T), decoupling can considerably improve the quality of spectra [28] (see Fig. 2). At higher field strength however, this approach will be of little help. The exact assignment of the resonances has to be made in tissue extracts. It has been shown that the extraction procedure usually does not affect the levels of mobile PMs and PDs [135].

## 5.3 *Phosphatidylinositol Turnover*

The intracellular biological effects of many hormones and neurotransmitters that are not able to enter the target cell themselves are transduced across the cell membrane by specific mechanisms that initiate intracellular signal cascades. One such fundamental transducing mechanism, which may be investigated in vivo by $^{31}$P NMR, is schematically depicted in Fig. 5. Upon binding of agonists to extracellular receptors, membrane bound phosphodiesterase (PDEase) is activated and cleavage of membrane phospholipids (phosphatidylinositol-bi-

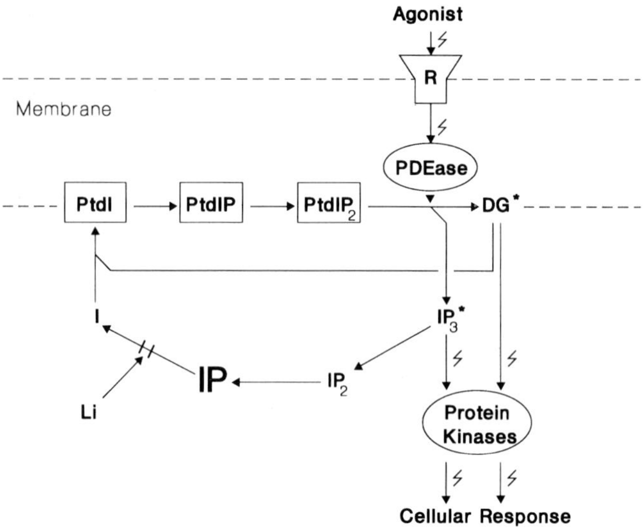


**Fig. 5.** Simplified scheme representing phosphatidylinositol turnover. Stimulation of an extracellular receptor by an agonist, which is unable to cross the cell membrane, activates membrane bound phosphodiesterase (PDEase) leading to cleavage of phosphatidylinositoldiphosphate ($PtdIP_2$), thereby producing inositol-1,4,5-triphosphate ($IP_3$) and 1,2-diacylglycerol (DG). These two 'second messengers' (indicated by an asterisk) then activate intracellular signal cascades leading to a cellular response. The 'second messengers' are inactivated and recycled to replenish the phosphatidylinositol (PdtI) pool (see text). The activity state of the cell correlates with the PdtI turnover. The concentration of the individual inositolphosphates, however, is too low to be detected by NMR. If inositol-1-phosphatase is inhibited with lithium (Li), inositol-1-phosphate (IP) accumulates to reach millimolar concentration, which may be detected with $^{31}P$ NMR. The signal appears in the PM region of the spectrum

phosphate = $PtdIP_2$) is induced, resulting in the intracellular formation of diacylglycerol (DG) and inositol triphosphate ($IP_3$). These two metabolites then function as 'second messengers' to activate intracellular signal cascades. $IP_3$ causes release of calcium, leading to activation of protein kinases, which then activate through phosphorylation, a variety of peptides and proteins, such as growth factors, ion channels, enzymes etc. These cellular responses will thus depend upon the levels of 'second messengers' [136]. In order to promptly terminate these cellular responses when the external signal (receptor agonist) is withdrawn, efficient degradative pathways that rapidly remove these 'second messengers', are present: $IP_3$ is inactivated by stepwise dephosphorylation to inositol (I), while DG is phosphorylated to form phosphatidic acid. Both inactivated 'second messengers' are finally recycled back by recombining to replenish the pool of phosphatidylinositol (PtdI).

Even if $^{31}P$ NMR would be sensitive enough to measure directly the in vivo levels of the phosphate containing intermediates of the PtdI cycle, no information concerning the extent of the cellular response could be deduced from such data, because the levels of these phosphates as discussed, are kept low and constant and therefore do not correlate with the activity of the transducing system. Instead, in analogy to the situation discussed for HEPs, the turnover of PtdI correlates with the activity state of the transducing system and therefore reflects the extent of the physiologically relevant cellular responses.

A measure of the PtdI turnover is the accumulation of inositol-1-phosphate (IP) after inhibition of inositol-1-phosphatase by $Li^+$ [137]. In vivo, a single subcutaneous dose of 10 mmol/kg LiCl gives rise to an increase of IP in various organs, e.g., a 40 fold increase was found after 24 h in the rat cerebral cortex [138]. This accumulation presumably represents the integrated total brain agonistic activity at receptors using PtdI as transducing system, as it is diminished by anaesthetics such as halothane or barbiturates. The acute effects of drugs acting on the central nervous system, e.g. cholinergic agonists, can also be revealed in vivo in a similar way. Following the pretreatment of a rat with a low dose of LiCl (3 mmol/kg), which by itself leads to only a modest (2–3 fold) increase of IP after 24 h, administration of pilocarpine (30 mg/kg) will cause a 40 fold increase of IP within 60 min. The accumulation of IP following administration of agonists can be prevented with corresponding *anta*gonists, e.g. atropine, but only occurs, if the inositol-1-phosphatase is inhibited by $Li^+$ [138].

The IP resonance occurs in the phosphomonester region of the $^{31}P$ spectrum. Renshaw et al. [139] demonstrated that in LiCl pretreated cats the PM resonance

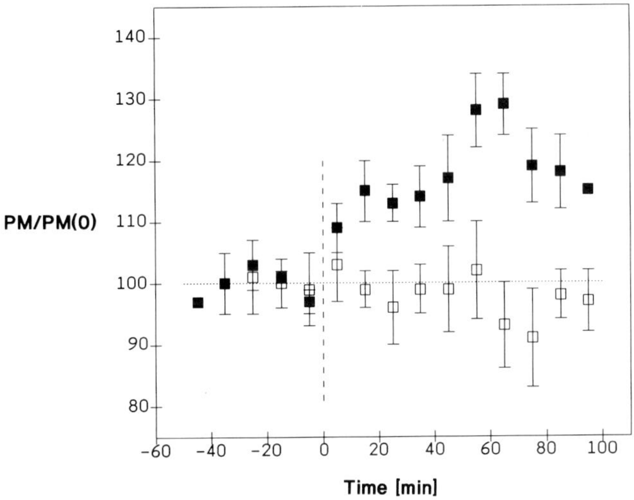


**Fig. 6.** Receptor stimulated formation of inositol-1-phosphate (IP) in rat brain. One group of rats (*full square*, N = 9) received a dose of 3 mmol/kg LiCl, 24 h before the experiment, the second group (*open squares*, *N* = 8) had no pretreatment. At the beginning of the experiment, four to five control spectra were recorded. The initial PM signals were identical in the two groups. At time t = 0, Pilocarpine, a cholinergic agonist, was injected intraperitoneally at a dose of 30 mg/kg in both groups of animals. Pilocarpine administration led in the LiCl pretreated group to a 30% increase in the PM resonance, reaching the maximum 60 min after injection. No signal change was observed in the second group

increased by 35% following administration of pilocarpine. High resolution NMR spectra of brain extracts confirmed that this increase was in fact due to an increase of IP. We confirmed these results in the rat. Figure 6 shows the time course of the relative PM signal recorded from the cerebral cortex using a surface coil following injection of pilocarpine: Within 60 minutes a 30% increase is observed, but only in LiCl pretreated animals. Assuming that the normal PM concentration in brain is 3–4 mM [140], the 30% intensity increase caused by LiCl/pilocarpine translates to a concentration change of about 1 mM. This is in good agreement with the values reported (10 mmol/kg dry weight) using conventional gas chromatographic methods [138].

The limitations of $^{31}$P NMR to measure PtdI turnover have to be weighed against the obvious advantages. Since only an increase of the PM resonance can be observed, the specificity of the method is low. In addition, due to the inherent sensitivity constraints, maximal receptor stimulation is required, which may rapidly lead to exhaustion of the PtdI pool and hence, disturb the physiological homeostasis in a too drastic way. On the other hand, the non-invasiveness of in vivo NMR methods offers numerous advantages for pharmacological studies.

## 6 Conclusions and Perspectives

In vivo NMR allows to study processes of fundamental biological importance in an undisturbed environment, opening a large field of unique applications, both in experimental research and in clinical practice. $^{31}$P NMR is a major domain of in vivo NMR, contributing to our understanding of biological processes involving $^{31}$P containing compounds. The inherent specificity of the method is frequently lost under in vivo conditions. The assignment of resonances and the unambiguous interpretation of the spectra may then become difficult, but can be improved by using editing techniques, proton decoupling [28] or correlated spectroscopy (Navon et al., this volume). Different NMR techniques may produce irregular results and lead to ambiguous conclusions. These points have been discussed extensively by Bottomley et al. [141].

The main efforts in $^{31}$P NMR are at present directed towards experimental studies with animal models of human disorders. By understanding the biochemical basis of different pathologies one may eventually devise strategies for new treatments. The effects of drugs and ultimately the success of a therapy may also be monitored using non-invasive NMR techniques.

The principle conclusion from studies trying to relate energy metabolism to pathological conditions and drug effects is that the levels of HEP's, in particular ATP, are well regulated and hence, are insensitive indicators of the subtle, but nevertheless important biological changes occurring in pathological tissue. This may in fact explain the rather limited success of $^{31}P$ NMR so far in clinical diagnosis. On the other hand, $^{31}P$ NMR offers also the possibility to non-invasively determine the turnover of HEP's, which correlate more dynamically with organ function than the steady-state tissue HEP levels [73, 116, 107].

Probably the most relevant clinical contribution of $^{31}P$ NMR has been made in studies of hereditary myopathies [103]. Other rather successful examples were the establishment of metabolite related prognostic indexes for intermittant claudication [101], birth asphyxia [95, 96, 126], or brain maturation [123]. Attempts to use $^{31}P$ NMR for the characterization of tumors (e.g., differentiation benign/malignant), have not been successful up to now. Whether this method will be of value in monitoring the response of tumors to therapy, is still an open question. Early changes in the spectral and thus metabolic patterns following the initiation of radiation or chemotherapy have been reported [132, 142–145]. Whether such findings are predictive for the success of the therapy remains to be shown. Interesting new applications with potential clinical relevance are $^{31}P$ NMR studies of the phospholipid metabolism, since this may lead to a better characterization of neurodegenerative diseases. The value of $^{31}P$ NMR for the study of PdtI turnover, i.e., signal transduction and 'second messenger' formation in the cell, and the application of saturation transfer methods for diagnosis and better understanding of human disorders remains to be demonstrated.

*Acknowledgement*: The authors are greatful to Dr. Joachim Seelig and Dr. Hans Widmer for critically reading the manuscript.

## 7 References

1. Moon RB, Richards JH (1973) J Biol Chem 284: 7276
2. Hoult DI, Busby SJW, Gadian DG, Radda GK, Richards RE, Seeley PJ (1974) Nature 252: 285
3. Gordon RE, Hanley PE, Shaw D (1982) Progr Nucl Magn Res Spectr 15: 1
4. Ackerman JJH, Grove TH, Wong GG, Gadian DG, Radda GK (1980) Nature 283: 167
5. Ordidge RG, Connelly A, Lohman JAB (1986) J Magn Reson 66: 283
6. Moudsley A, Hilal SK, Simon HE, Wittekoek S (1984) Radiology 153: 745
7. Weiner MW, Hetherington H, Hubesch B, Kraczmar G, Massie B, Maudsley AA, Meyerhoff DJ, Sappey-Marinier D, Schaefer S, Twieg DB, Matson GB (1989) NMR Biomed 2: 290
8. Maudsley AA, Twieg TB, Sappey-Marinier D, Hubesch B, Hugg JW, Matson GB, Weiner MW (1990) Magn Reson Med 14: 415
9. Bottomley PA, Foster TB, Darrow RD (1984) J Magn Reson 59: 338
10. Robitaille PM, Merkle H, Sublett E, Hendrich K, Lew B, Path G, From AHL, Boche RJ, GarwoodM, Ugurbil K (1989) Magn Reson Med 10: 14
11. Brown TR, Kincaid BM, Ugurbil K (1982) Proc Natl Acad Sci, USA 79: 3532
12. Haselgrove JC, Subramanian VH, Leigh JS, Gyulai L, Chance B (1983) Science 220: 1170
13. Koretsky AP, Wang S, Murphy-Boesch J, Klein MP, James TL, Weiner MW (1983) Proc Natl Acad Sci, USA 80: 7491

14. Gordon RE, Timms WE (1982) J Magn Reson 46: 322
15. Houkin K, Nakada T, Suzuki N, Kwee IL (1989) Magn Reson Med 12: 364
16. Silver MS, Jospeh RI, Hoult DI (1984) J Magn Reson 59: 347
17. Ugurbil K, Garwood M, Rath AR (1988) J Magn Reson 80: 448
18. Alderman DW, Grant DM (1979) J Magn Reson 36: 447
19. Leroy-Willig A, Darrasse L, Taquin J, Sauzade M (1985) Magn Reson Med 2: 20
20. Cross TA, Mueller S, Aue WP (1985) J Magn Reson 62: 87
21. Hayes CE, Edelstein WA, Schenk JF, Mueller OM, Eash M (1985) J Magn Reson 63: 622
22. Eveloch JL, Ewy CS, Siegfried BA, Ackerman JJH (1985) Magn Reson Med 2: 410
23. Mathur-DeVre M, Maerschalk C, Mattrey RF (1990) Magn. Reson Imag 8: 691
24. Aue WP, Mueller S, Cross TA, Seelig J (1984) J Magn Reson 56: 350
25. Hoult DI, Richards RE (1976) J Magn Reson 24: 71
26. Buchtal SD, Thoma WJ, Taylor JS, Nelson SJ, Brown TR (1989) NMR Biomed 2: 298
27. Bachert-Baumann P, Ermark F, Zabel HJ, Sauter R, Semmler W, Lorenz WJ (1990) Magn Reson Med 15: 165
28. Luyten PR, Bruntink G, Sloff FM, Vermeulen JWAH, van der Heijden JI, Hollander JA, Heershap A (1989) NMR Biomed 1: 177
29. Cohen SM, Burt CT (1977) Proc Natl Acad Sci, USA 74: 4271
30. Barany M, Glonek T (1984) In: Gorenstein D (ed) Phosphorus-31 NMR, Academic San Diego, chap 17
31. Strear CM, Bolinger L, Leigh JS (1989) Proc 8th Ann Meeting Soc Magn Reson Med, Amsterdam p 187
32. Shaka AJ, Keeler J, Fremkiel T, Freeman R (1983) J Magn Reson 52: 335
33. Brindle KM, Smith MB, Rajagopalan B, Radda GK (1985) J Magn Reson 61: 559
34. Ackerman JJH, Eveloch JL, Berkowitz BA, Kichura GM, Deuel RK, Lown KS (1984) J Magn Reson 56: 318
35. Hsieh PS, Balaban RS (1987) J Magn Reson 74: 574
36. Mora BN, Narashimhan PT, Barker PB, Ross BD, Allman JM (1990) Proc 9th Ann Meeting Soc Magn Reson Med, New York p 1020
37. Pettegrew JW, Kopp SJ, Dadok J, Minshew NJ, Feliksik JM, Glonek T, Cohen MM (1986) J Magn Reson 67: 443
38. Gonzales-Mendez R, Litt L, Koretsky AP, von Colditz J, Weiner MW, James TL (1984) J Magn Reson 57: 526
39. Murphy EJ, Rajagopalan B, Brindle KM, Radda GK (1989) Magn Reson Med 12: 282
40. Campbell ID, Dobson CM, Williams RJP, Xavier AV (1973) J Magn Reson 11: 172
41. Campbell ID, Dobson CN, Williams RJP, Wright PE (1975) FEBS Lett 57: 961
42. Brown FF, Campbell ID (1980) Philos Trans Roy Soc London Ser B 289: 395
43. Morris GA, Freeman R (1978) J Magn Reson 29: 433
44. Thoma WJ, Henderson LM, Ugurbil K (1985) J Magn Reson 61: 141
45. Kushnir T, Kaplan O, Askenasy N, Navon G (1989) Magn Reson Med 10: 119
46. Carrington A, McLachlan AD (1969) 'Introduction to Magnetic Resonance', Harper & Row, New York p 204
47. Pettegrew JW, Withers G, Panchalingam K, Post JFM (1988) Magn Res Imag 6: 135
48. Bailey IA, Williams SR, Radda GK, Gadian DG (1981) J Biochem 196: 171.
49. Naruse S, Takada S, Koizuka I, Watari H (1983) J Japn Physiol 33: 19
50. Benabid AL, Decorps M, Remy C, LeBas JF, Confort S, Leviel JL (1987) J Neurochem 48: 804
51. Seo Y, Murakami M, Watari H, Imai Y, Yoshizaki K, Nishikawa H, Morimoto T (1983) J Biochem 94: 729
52. Jacobus WE, Pores IH, Lukas SK, Kallman CH, Weisfeldt ML, Flaherty JT (1982) In: Nucitelli R, Deamer DW (eds) 'Intracellular pH: its measurement, regulation and utilization in cellular functions' Allan R. Liss, New York p 537
53. Garlick PB, Radda GK, Seeley PJ (1979) Biochem J 184: 547
54. Gard JK, Kichura GM, Ackerman JJH, Eisemberg JD, Billadello JJ, Sobel BE, Gross RW (1985) J Biophys 48: 803
55. Petroff OAC, Prichard JW, Behar KL, Alger JR, den Hollander JA, Shulman RG (1985) Neurology 35: 781
56. Mitsumori F (1985) J Biol Chem 97: 1551
57. Gupta RK, Gupta P, Yushok WD, Rose ZD (1983) Physiol Chem Phys Med NMR 15: 266
58. Wu ST, Pieper GM, Salhany JM, Eliot RS (1981) Biochemistry 20: 1399
59. Shulman RG, Brown TR, Ugurbil K, Ogawa S, Cohen SM, Hollander JA (1979) Science 205: 160

60. Griffiths JR, Stevens AN, Iles RA, Gordon RE, Shaw D (1981) Biosci Rep 1: 319
61. Deuel RK, Yue GM, Sherman WR, Schickner DJ, Ackerman JJH (1985) Science 228: 1329
62. Koga K, Miura I (1988) Biochem Biophys Res Comm 157: 1258
63. Clarke K, Nedelec JF, Mills PA, Ingwall JS (1988) Proc 7th Ann Meeting Soc Magn Reson Med, San Francisco, p 177
64. Cohn M, Hughes TR (1962) J Biol Chem 237: 176
65. Gupta RK, Benovic JL, Rose ZB (1978) J Biol Chem 253: 6172
66. Gupta RK, Yushok WD (1980) Proc Natl Acad Sci USA 77: 2478
67. Gupta RK, Moore RD (1980) J Biol Chem 255: 3987
68. Vink R, McIntosh TK, Demediuk P, Weiner MW, Faden AI (1988) J Biol Chem 263: 757
69. Vink R, Yum SW, Lemke M, Demediuk P, Faden AI (1989) Barin Res 490: 144
70. Cohen SM, Burt CT (1977) Proc Natl Acad Sci, USA 74: 4271
71. Clarke K, Balschi JA, Neubauer S, Kleber AG, Springer CS, Smith TW, Ingwall JS (1989) Proc 8th Ann Meeting Soc Magn Reson Med, Amsterdam p 665
72. Tofts PS, Wray S (1988) NMR Biomed 1: 1
73. Bittl JA, DeLayre J, Ingwall JS (1987) Biochem 26: 6083
74. Roth K, Hubesch B, Meyerhoff DJ, Naruse S, Gober JR, Lawry TJ, Boska MD, Matson GB, Weiner MW (1989) J Magn Reson 81: 299
75. Rehncrona S, Rosen I, Siesjo BK (1981) J Cereb Blood Flow Metabol 1: 297
76. Brindle KM, Blackledge MJ, Challiss RAJ, Radda GK (1989) Biochem 28: 4887
77. Malloy CR, Cunningham CC, Radda GK (1986) Biochem Biophys Acta 885: 1
78. Stromski ME, Cooper K, Thulin G, Gaudio KM, Siegel NJ, Shulman RJ (1986) Proc Natl Acad Sci, USA 83: 6142
79. Ross B, Freeman D, Chan L (1986) Kidn Internat 29: 131
80. Shine NR, Xuan JA, Koretsky AP, Weiner MW (1987) Magn Res Med 4: 244
81. Bottomley PA, Kogure K, Namon R, Alonso OF (1982) Magn Reson Imag 1: 81
82. Delpy DT, Gordon RE, Hope PL, Parker D, Reynolds EOR, Shaw D, Whitehead MD (1982) Pediatrics 70: 310
83. Thulborn KR, duBoulay GH, Duchen LW, Radda GK (1982) J Cereb Blood Flow Metab 2: 299
84. Naruse S, Horikawa Y, Tanaka C, Hirakawa K, Nishikawa H, Watari H (1984) Neurol Res 6: 169
85. Horikawa Y, Naruse S, Hirakawa K, Tanaka C, Nishikawa H, Watari H (1985) J Cereb Blood Flow Metab 5: 235
86. Bottomley PA, Drayer BP, Smith SL (1986) Radiology 160: 763
87. Gadian DG, Frackowiak RSJ, Crockard HA, Proctor E, Allen K, Williams SR, Russell RWR (1987) J Cereb Blood Flow Metab 7: 199
88. Bolas N, Rajagopalan B, Mitsumori F, Radda GK (1987) Stroke 5: 608
89. Prichard JW, Alger JR, Behar KL, Petroff OAC, Shulman RG (1983) Proc Natl Acad Sci, USA 80: 2748
90. Ishige N, Pitts LH, Pogliani L, Hashimoto T, Nishimura MC, Bartkowski HM, James TL (1987) Neurosurg 20: 854
91. Gonzales-Mendez R, McNeill A, Gregory G, Wall SD, Gooding CA, Litt L, James TL (1985) J Cereb Blood Flow Metab 5: 512
92. Chopp M, Helpern JA, Ewing JR, Welch KMA (1984) Magn Reson Imag 2: 329
93. Kopp SJ, Kriegelstein J, Freidank A, Rachman A, Seibert A, Cohen MM (1984) J Neurochem 43: 1716
94. Alger JR, Brunetti A, Nagashima G, Hossmann KA (1989) J Cereb Blood Flow Metab 9: 506
95. Cady EB, Dawson MJ, Hope PL, Tofts PS, Costello AML, Delpy DT, Reynolds EOR, Wilkie DR (1983) Lancet 1: 1059
96. Hamilton PA, Cady EB, Wyatt JS, Hope PL, Delpy DT, Reynolds EOR (1986) Lancet 1: 1242
97. Petroff OAC, Prichard JW, Behar KL, Alger JR, Shulman RG (1984) Ann Neurol 16: 169
98. Nunnally RL, Bottomley PA (1981) Science 211: 177
99. Bottomley PA, Herfkens RJ, Smith LS, Brazzamano S, Blinder R, Hedlund LW, Swain JL, Redington RW (1985) Proc Natl Acad Sci, USA 82: 8747
100. Robitaille PM, Lew B, Merkle H, Sublett E, Lindstrom P, From AHL, Garwood M, Bache RJ, Ugurbil K (1989) Magn Reson Med 10: 108
101. Keller U, Oberhaensli R, Huber P, Widmer LK, Aue WP, Hassink RI, Mueller S, Seelig J (1985) Eur J Clin Invest 15: 382
102. Williams DM, Fencil L, Chenevert TL (1990) Radiol 175: 381
103. Taylor DJ, Bore PJ, Styles P, Gadian DG, Radda GK (1983) Mol Biol Med 1: 1983
104. Radda GK (1986) Science 233: 640

105. Sauter A, Rudin M (1987) Magn Res Med 4: 1
106. Thulborn KR (1981) Biochem Soc Trans 9: 237
107. Rudin M, Sauter A (1989) J Pharm Exp Therap 251: 700
108. Allen A, Busza AL, Crockard HA, Frackowiak RSJ, Gadian DG, Proctor E, Russel RWR, Williams SR (1988) J Cereb Blood Flow Metab 8: 816
109. Bruehn H, Frahm J, Gyngell ML, Merboldt KD, Haenicke W, Sauter R (1989) Magn Reson Med 9: 126
110. Walsh KMA, Gross B, Licht J, Levine SR, Glasberg M, Smith MB, Helpern JA, Bueri J, Gorell J (1988) In: Apple S (ed) Current Neurology, Vol 8, Year Book Publishers, Chicago p 295
111. Michenfelder JD, Sundt TM (1971) Stroke 2: 319
112. Rudin M, Sauter A (1989) Proc 8th Ann Meeting Soc Magn Reson Med, Amsterdam p 489
113. Williams SR, Proctor E, Allen K, Gadian DG, Crockard HA (1988) Magn Reson Med 7: 425
114. Behar KL, den Hollander JA, Stromski ME, Ogino T, Shulman RG, Petroff OAC, Prichard JW (1983) Proc Natl Acad Sci, USA 80: 4945
115. Sokoloff L (1982) In: 'Advances in Neurochemistry', Vol. 4, (Agranoff BW, Apison MH, eds), Plenum, New York p 1
116. Bittl JA, Ingwall JS (1985) J Biol Chem 260: 3512
117. Pettegrew JW, Withers G, Panchalingam K, Post JF (1987) J Neural Transmiss Suppl 24: 261
118. Gyulai L, Bolinger L, Leigh JS, Barlow C, Chance B (1984) FEBS Lett 178: 137
119. Pettegrew JW, Kopp SJ, Dadok J, Minshew NJ, Feliksik JM, Glonek T, Cohen MM (1986) J Magn Reson 67: 443
120. Glonek T, Kopp SJ, Kot E, Pettergrew JW, Harrison WH, Cohen MM (1982) J Neurochem 39: 1210
121. Cerdan S, Subramanian VH, Hilberman M, Cone J, Egan J, Chance B, Williamson JR (1986) Magn Reson Med 3: 432
122. Pettegrew JW, Kopp SJ, Minshew NJ, Glonek T, Feliksik JM, Tow JP, Cohen MM (1987) J Neuropathol Exp Neurol 46: 419
123. Boesch C, Gruetter R, Martin E, Duc G, Wuethrich K (1989) Radiology 172: 197
124. Tofts P, Wray S (1985) J Physiol 359: 417
125. Eveloch JL, Keller NA, Corbett YH (1987) Cacner Res 47: 3396
126. Hope PL, Cady EB, Tofts PS, Hamilton PA, Costello AML, Delpy DT, Chu A, Reynolds EOR, Wilkie DR (1984) Lancet 2: 366
127. Johnson MA, Pennock JM, Bydder GM, ... (1983) J Am Radiol 141: 1005
128. Martin E, Kikinis R, Zuerrer M, ... (1988) J Comput Assist Tomogr 12: 917
129. Brown GG, Levine SR, Gorell JW, Pettegrew JW, Gdowski JW, Bueri JA, Helpern JA, Welch KMA (1989) Neurol 39: 1423
130. Sostman HD, Armitage IM, Fischer JJ (1984) Magn Reson Med 2: 265
131. Glickson JD, Evanochko WT, Sakai TT (1986) In: Gupta R (ed), NMR Spectroscopy of Cells and Organism, CRC Press
132. Evanochko WT, Ng TC, Glickson JD (1984) Magn Reson Med 1: 508
133. Rudin M, Tolcsvai L (1989) unpublished
134. Luyten P, Heindel W, Marien WAJH, Herholz K, Heiss WD, Friedmann G, den Hollander JA (1990) Proc 9th Ann Meeting Soc Magn Reson Med (New York) p 103
135. Cohen MM, Pettegrew JW, Kopp SJ, Minshew N, Glonek T (1984) Neurochem Res 9: 785
136. Berridge MJ (1984) Bichem J 220: 345
137. Hallcher LM, Sherman WR (1980) J Biol Chem 255: 10896
138. Sherman WR, Gish BG, Honchar MP, Munsell LY (1986) Fed Proc 45: 11
139. Renshaw PF, Schnall MD, Leigh JS (1987) Magn Reson Med 4: 221
140. Bethesda MD, Altman PL, Dittmer DS (eds) Biological Data Handbook Feder Am Soc Exptl Biol p 1211
141. Bottomley PA, Hardy CJ, Roemer PB, Weiss RG (1989) NMR Biomed 2: 284
142. Evanochko WT, Ng TC, Lilly MB, Lawson AJ, Corbett TH, Durant JR, Glickson JD (1983) Proc Natl Acad Sci, USA 80: 334
143. Segebarth CM, Baleriaux DF, Arnold DL, Luyten P, Hollander JA (1987) Radiology 165: 215
144. Ross B, Helsper JT, Cox IJ, Young IR, Kempf R, Makepeace A, Pennock J (1987) Arch Surg 122: 1464
145. Semmler W, Gademann G, Bachert-Baumann P, Zabel HJ, Lorentz WJ, van Kaik G (1988) Radiology 166: 533
146. Schenker K, Link J, Brunner P (1991) Proc 8th Ann Meeting European Soc Magn Res Med Biol, ESMRMB, Zürich, p 272

# Author Index Volumes 21–28